Lecture Notes in Computer Science 651

Edited by G. Goos and J. Hartmanis

Advisory Board: W. Brauer   D. Gries   J. Stoer

Ron Koymans

Specifying Message Passing and Time-Critical Systems with Temporal Logic

Springer-Verlag

Berlin Heidelberg New York
London Paris Tokyo
Hong Kong Barcelona
Budapest

Series Editors

Gerhard Goos
Universität Karlsruhe
Postfach 69 80
Vincenz-Priessnitz-Straße 1
W-7500 Karlsruhe, FRG

Juris Hartmanis
Cornell University
Department of Computer Science
4130 Upson Hall
Ithaca, NY 14853, USA

Autor

Ron Koymans
Philips Research Laboratories, P.O.B. 80 000
NL-5600 JA Eindhoven, The Netherlands

CR Subject Classification (1991): F.3.1, F.4.1, J.7

ISBN 3-540-56283-4 Springer-Verlag Berlin Heidelberg New York
ISBN 0-387-56283-4 Springer-Verlag New York Berlin Heidelberg

© Springer-Verlag Berlin Heidelberg 1992
Printed in Germany

Typesetting: Camera ready by author/editor
Printing and binding: Druckhaus Beltz, Hemsbach/Bergstr.
45/3140-543210 - Printed on acid-free paper

List of Figures

Πάντα χωρεῖ καὶ οὐδὲν μένει (Herakleitos, ±500 B.C.)

Preface

This monograph is an updated and extended version of my Ph.D. thesis [Koy 89]. It is concerned with the application of temporal logic to the areas of message passing and time-critical systems. Apart from the practical use of temporal logic for these two application domains this monograph also incorporates pure fundamental studies on temporal logic. This duality may stem from my education: after studying (mathematical) logic I went on to finish my study in computer science. This is reflected in my main research interest: putting (mathematical) theory into (computer science) practice. Some readers may not be interested in the combination of theory and practice. To those interested mainly in theoretical results I can recommend reading Chapters 3 and 4 and sections 5.1, 5.2, 5.4, 6.1, 6.2 and 6.4. Readers interested more in practical issues could read Chapters 2, 5 and 6, and the following preliminaries from Chapters 3 and 4: section 3.1, section 3.2 till Definition 3.2.24, the definitions of **until** and **since** in section 3.3, section 3.4, section 4.1, section 4.2 till after Proposition 4.2.10, and section 4.4.

Writing my thesis would not have been possible without the help and support of many people. I thank Willem-Paul de Roever and Jan Vytopil for their stimulation and guidance in the last decade. Thanks go to Amir Pnueli who provided numerous suggestions for improvement of my work on many occasions and to Johan van Benthem who had a major influence on Chapter 4. From May 1984 till May 1988 I was involved in the Dutch National Concurrency Project (Dutch acronym LPC) which was supported by the Foundation for Computer Science Research in the Netherlands (SION) with financial aid from the Netherlands Organization for Scientific Research (NWO, formerly ZWO). Jaco de Bakker, Grzegorz Rozenberg and Peter van Emde Boas are thanked for their useful criticisms at several presentations of my work. My former colleagues in the theoretical computer science group of the Eindhoven University of Technology were of great help in many discussions and also contributed in creating a pleasant working atmosphere. Karst has given invaluable assistance through all my studies since my childhood. I am indebted to my parents for their sympathy and encouragement. Special thanks are due to Letty for providing the necessary support.

Eindhoven, May 1992   Ron Koymans

Contents

Chapter 1

Introduction

This monograph is concerned with the development of specification methods that are applicable in the areas of message passing and time-critical systems. The motivation to study these application areas stems from their importance in practice:

- message passing is one of the most important means of interprocess communication in distributed systems, either on a high level (e.g. in telecommunication applications where programming could be done in a high-level concurrent language with asynchronous message passing such as CHILL [CHILL 80]) or on a lower level (such as in implementations of synchronous languages for distributed computing like Ada [Ada 83]),

- among the growing number of real-time applications there are some highly critical systems such as computer controlled chemical plants and nuclear power stations.

Because many of these systems belong to the most complex ever developed, adequate specification methods for them are of vital importance, a claim that is also supported by experience from practice: actual builders of systems see real-time as the most crucial area in which formal support is necessary.

This monograph reports on the application of temporal logic as a formalism for reasoning about message passing and time-critical systems. Such an enterprise was motivated by noticing that temporal logic had been applied very successfully for the specification and verification of a wide variety of systems, ranging from parallel programs (see e.g. [MP 82],[MP 83a],[MP 83b]) via communication protocols (see e.g. [HO 83]) to hardware/VLSI applications (see e.g. [Mos 83]). However, in the areas of message passing systems and of time-critical systems its application has been less successful. For message passing systems this has a technical reason: it can be shown that many classes of message passing systems cannot be specified with standard temporal logic. Since time-critical systems heavily involve quantitative temporal requirements and standard temporal logic is concerned only with reasoning about qualitative temporal issues, the inaptness of its application to time-critical systems is obvious. This monograph shows how standard temporal logic can still be used for the specification of message passing systems by introducing the additional assumption that incoming messages are uniquely identifiable and it develops a special temporal logic, called metric temporal logic, for reasoning about quantitative temporal properties. The main application area of this monograph can be found in the field of distributed real-time systems where message passing and time-critical features are combined.

Before one can use an established mathematical theory in new application areas, the fundamentals of this theory should be reinvestigated in the light of the peculiarities of these application areas and the objectives one strives to achieve. In fact, the development of a theory for a certain application area should ideally go hand in hand with checking (e.g. by means of paradigmatic cases) whether the theory works out in practice as intended. In our case, apart from undertaking theoretical studies involving possibilities to apply temporal logic in the areas of message passing and time-critical systems, we check the results against several examples taken from these application areas and against certain objectives one would like a specification method to have.

Historically, these ideas emerged in their preliminary form when the author was working in a project developing a digital telephone switching system at Philips Telecommunication Industries (Hilversum, the Netherlands) from September 1982 till June 1983. In telecommunication systems message passing aspects and time-critical aspects are combined, for instance in a time-out for the acknowledgment of a message when unreliable transmission media are involved (besides, current communication technology usually involves complex real-time software). The research in this period resulted (see [KVR 83]) in an axiomatic semantics for the real-time communication fragment of the concurrent programming language CHILL (see [CHILL 80]). After this practice period at Philips, the author was employed at the University of Nijmegen (from July 1983 till May 1984) before getting involved in the Dutch National Concurrency Project (acronym LPC) first at the University of Nijmegen (from

May 1984 till August 1985) and subsequently at the Eindhoven University of Technology (from August 1985 till May 1988). In this period the first variants of a temporal logic for reasoning about real-time properties, called real-time temporal logic, were developed and tested by means of examples from practice (see [KR 85]). For the specification of message passing we introduced the assumption that the incoming messages could be uniquely identified. We come back on this assumption below.

Another major research topic in this period, not reported upon in this monograph, was the work on a compositional semantics for real-time distributed computing taken up from September 1983 onwards. This research effort resulted (see [KSRGA 85]) in a denotational semantics for real-time distributed computing that is compositional in the context of process naming and nested parallelism. This semantics is based on a new class of real-time computation models varying from the interleaving model to the maximal parallelism model. These results were an important pillar for participation of the theoretical computer science group of the Eindhoven University of Technology in ESPRIT project 937: Debugging and Specification of Ada Real-Time Embedded Systems (DESCARTES).

The collaboration in the DESCARTES-project led to three more papers involving the specification of real-time systems: the first ([KKZ 87]) about another application of real-time temporal logic, in this case to the paradigms of real-time investigated in this project, the second ([KKZ 88]) about paradigms and a classification of real-time systems together with an informal account on requirements of a specification language for real-time properties and the third ([KKZ 89]) about a formal framework for treating and comparing requirements of a specification language. In these years the chosen way of using standard temporal logic for the specification of message passing systems (using the unique identification assumption) was supported by strengthening theoretical results about the (in)expressiveness of temporal logics for characterizing certain classes of message passing systems (see [Koy 87]). These results imply that message passing systems can only be specified using very strong logics (unless unique identification is assumed). After recognizing that the nomenclature real-time temporal logic was not fully justified since this logic was developed especially for reasoning about quantitative temporal properties only partially dealing with other important features of real-time systems such as reliability and performance issues, the logic was renamed to metric temporal logic and its application domain to the more general class of time-critical systems (nevertheless, real-time systems still remain the most prominent representatives of that class). However, metric temporal logic is not just another name for real-time temporal logic. Together with the renaming a theoretical study was undertaken regarding the fundamental principles underlying this logic. This resulted in an orientation towards the way temporal logic had been studied by philosophers for decades (in philosophy temporal logic is often called tense logic). Investigations of metric temporal logic about the interplay of qualitative and quantitative operators led to an interesting additional operator for modal and temporal logic which enables several formerly inexpressible natural assumptions about time to be expressed in the logic.

This monograph is organized as follows. Chapter 2 deals with requirements for a general specification language. After treating the embedding of a system in its environment with the interface in between it investigates how the behavior of a system should be specified.

Chapters 3 and 4 deal with several variants of modal and temporal logic. First, Chapter 3 gives a short recapitulation of the basics of modal and temporal logic subdivided in classical modal and temporal logic (as studied by philosophers for decades, see e.g. [Pri 67]), temporal logics with **until** and **since** operators (as studied by Kamp and Stavi), and temporal logics used in computer science. Then, Chapter 4 extends classical modal and temporal logic with an additional modal/temporal operator. The expressive power of the resulting logics and several other semantic issues are investigated, complete axiomatizations are given, and decidability is proven.

Chapters 5 and 6 introduce the application domains of our interest and look at ways to specify these with temporal logic. Chapter 5 concerns message passing systems. First we describe which systems we consider to be message passing systems and we specialize the requirements from Chapter 2 to the specification of these systems. Next we prove inexpressiveness results of temporal logics for the specification of message passing systems (it turns out that many classes of message passing systems cannot be specified in strong temporal logics) and show how these logical limitations may be overcome. We illustrate this with three specification examples among which is a hierarchical specification of a layered communication network and end with some conclusions. Chapter 6 concerns time-critical systems. First we describe the characteristics of such systems and specialize the requirements of Chapter 2 to the specification of these systems. Next we introduce our special temporal logic for reasoning about quantitative temporal properties called metric temporal logic (MTL for short). Metric temporal logic is then illustrated by a series of examples involving time-critical (and often also message passing) features such as time-out, a watchdog timer, the wait/delay statement of concurrent programming languages and an abstract transmission medium. We end Chapter 6 with some conclusions.

At last, Chapter 7 looks at the obtained results in retrospect, presents some conclusions, mentions related work and lists possibilities for future research.

Chapter 2

How to Specify

A whole book could be devoted to the topic of requirements of a general specification language. In this chapter we restrict ourselves to a small set of desirable properties for a specification language for general systems. In section 3 of Chapter 5 and section 3 of Chapter 6 we reconsider this topic for message passing, respectively time-critical systems. For a more extensive theoretical account on the subject of specification we refer the reader to [KKZ 89].

So, before we look at the systems of our special interest, viz. message passing systems and time-critical systems, and how to specify them in chapters 5 and 6, we first study the issue of specifying systems in general. To start with, one of the main characteristics of a system is that it does not work in isolation but exchanges information with its environment. So, each system can be viewed as being embedded in some environment consisting of the external sources and recipients of the data interchanged. The environment may consist of computer systems, but also physical processes and humans. Pictorially this may be represented as in Figure 2.1. In this figure the environment surrounds the system residing inside some boundary that demarcates the scope of responsibility of the system. This boundary between the system and its environment, formed by the collection of data elements interchanged between them, constitutes what we will call the (abstract) interface. The interface is all the environment sees of the system and the other way around. This use of the word interface relates to abstract entities and indicates only what kind of data is interchanged and should not be confused with the physical interface where it is indicated how this exchange is achieved physically (the RS-232 serial interface for data communication is a typical example). We will see an example of an abstract interface in section 2 of

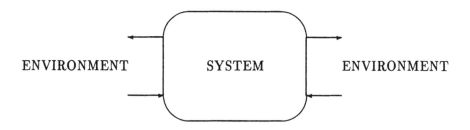

Figure 2.1: System in its Environment

Chapter 5 where we treat message passing systems.

A data element of the interface is a primitive entity that is by definition observable since it is interchanged between the system and the outside world. The data elements can be partitioned into two categories: state variables and events. A state variable is persistently present, that is, it has a value at each moment (for example a temperature sensor) while an event represents an entity that is intermittently present, that is, it occurs at discrete moments (e.g. the arrival of a message). When some occurrences of an event are not instantaneous but can have some duration (e.g. the transmission of a message over a communication link) we call it an action or extended event in order to distinguish it from an (instantaneous) event which only occurs at discrete points in time. Two events can be causally related to each other such as the response of the system to a stimulus from the environment. A causal relationship between events implies a temporal ordering of these events, but not the other way around. For example, a response can never occur before the corresponding stimulus. As an abstract view it may be helpful to allow simultaneity of causally related events but it should be remembered that this cannot be implemented because that would involve the possibility of infinite speed.

When specifying systems in general this should include a specification of the interface between the system and its environment. Although it is usually sufficient to give the intuitive interpretation of the data elements involved (together with attributes such as responsibility and the direction of the information flow) this is an essential part of the specification.

Next the behavior of the system in its environment has to be specified. In order not to restrict oneself a priori to a certain set of implementation possibilities, such a specification should only specify the requirements put on the system when operating in a certain environment and not any design details relating to the internal operation of the system (because such details would suggest certain ways to achieve the required behavior and hence would be implementation biased). For the representation of a system as embedded in its environment this involves viewing the system from the outside, as a black box. Such a viewpoint leads to the notion of observable equivalence: systems that behave the same as observed only from the outside are considered equivalent (although they may differ considerably internally). Hence, the specification of the required behavior should have exactly that level of abstraction which differentiates between unwanted and allowed implementations, that is, it should be sufficiently concrete to rule out unwanted implementations and sufficiently abstract to cater for all (allowed) possible implementations. This notion of the right level of abstraction is a semantic one since it is based on the semantic relation of satisfaction between an implementation and a specification (in the context of semantics it is often referred to by 'fully abstractness'). However, this semantic notion of the right level of abstraction is not sufficient for our purpose. We intend the specification of behavior to be completely free of any implementation bias whatsoever, implying that not even syntactically implementations should be suggested. This relates to a common way to achieve the right level of abstraction semantically by hiding internal variables introduced in the specification by means of some abstraction mechanism (usually connected to some form of existential quantification). For example, consider a specification that uses an internal variable pc representing a program counter. This variable is clearly implementation biased, but by prefixing the specification with $\exists\, pc$ this internal variable has become semantically invisible. The result is that the *meaning* of the specification indeed gives the desired set of implementation possibilities but that the *form* of the specification suggests the use of certain extra internal variables. We intend to avoid such a syntactical implementation bias by demanding that the specification is phrased only in terms of the elements of the interface (as observed above these correspond precisely to the observable entities). We will call a specification without any implementation bias (neither syntactically nor semantically) syntactically abstract. This notion of syntactical abstractness is also briefly touched upon in [Pnu 86] in the context of compositionality. When one classifies specification languages as being descriptive (describing which behavior is required) or prescriptive (describing how the desired behavior can be achieved) it will be clear from the above that syntactical abstractness favors the descriptive ones.

Of course, our specification language should be formal to ensure rigorous analysis and verification of desired properties. Further advantages of a formal approach include:

- in the process of formalization ambiguities, omissions and contradictions in the informal requirements can be detected,

- a formally verified part can be embedded with more confidence that it will function correctly (the formal model leads to enhanced reliability),

- the formal model can be a foundation for (partly) automated design methods and tools such as simulators,

- several designs can be compared.

Two further desirable properties of a general specification language are in our opinion:

1. conformity: similar systems have similar specifications,

2. uniformity: the specification method is based on a single formalism covering all aspects of a specification.

In section 5 of Chapter 5 we will review these properties in the context of message passing systems. This is done by contrasting specification methods that lack one or more of these properties with an approach that does incorporate all of them. In this process we also indicate which price has to be paid for attaining these properties.

The next pair of requirements for a general specification language is simplicity versus adequacy. On one hand simplicity increases understandability and usability, on the other hand the language should be powerful enough to describe all desired properties. For complex systems these two requirements are in conflict. In such a case the problem consists of finding a language that is as simple as possible but still has sufficient expressive power.

If the specification language is also used for the design of complex systems, this can only be done in a structured fashion by using several layers of abstraction. In this context it is essential that the method supports both top-down and bottom-up development techniques. This is tightly connected with the notions of compositionality and modularity (see e.g. [Zwi 88],[Pan 88],[Jon 87]). For top-down development the method should

be compositional, that is, to a chosen decomposition of the system there is
always a corresponding decomposition of the specification. For bottom-up
development modularity is essential, that is, it should always be possible
to combine given components in a way that all properties of the resulting
combination can be derived from the specifications of these components.

Apart from the above more theoretical requirements it makes sense to
include also requirements with respect to the practical usability of a specification language. Typical examples of such requirements are easy understandability, easy readability (by using a suitable representation) and easy
maintainability. For more information on this topic we refer the reader to
an extensive survey of such requirements that has been undertaken in RACE
project 2039 'SPECS'.

Chapter 3

A Review of Modal and Temporal Logic

3.1 Introduction

In this chapter we give a brief overview of notions and results from modal and temporal logics used in philosophy and/or computer science that are needed as a background for later chapters.

We start in section 2 with the way modal and temporal logics have been used in philosophy since decades (see e.g. [Pri 67]). We describe of course the syntax and semantics of such logics and look at some issues of correspondence theory (see [Ben 84]), axiomatizations and decidability.

In section 3 we look at temporal logics with **until** and/or **since** operators as studied by Kamp (see [Kam 68]) and Stavi (see [Sta 79],[Gab 81]). Apart from syntax and semantics of these logics we look at expressive completeness results.

At last, section 4 looks at some specialized temporal logics used in computer science such as linear time, branching time and interval temporal logics and how such logics can be used as a specification language.

3.2 Classical Modal and Temporal Logic

In this section we recapitulate the basics of propositional modal and temporal logic. In this and the next chapter we will use the following notational conventions. By $\varphi, \varphi_1, \ldots, \psi, \psi_1, \ldots, \chi, \chi_1, \ldots$ we denote formulas and by $\Phi, \Phi_1, \ldots, \Psi, \Psi_1, \ldots$ sets of formulas. We start out from a propositional language containing proposition letters $(p, p_1, p_2, \ldots, q, \ldots)$, two propositional constants $\bot$ (falsum) and $\top$ (verum), and the boolean operators $\neg$ (not), $\wedge$ (and), $\vee$ (or), $\rightarrow$ (if ... then ...) and $\leftrightarrow$ (if and only if). As a basis for this language we take $\{\rightarrow, \bot\}$ from which the other constant and other boolean operators can be defined as usual:

$$
\begin{aligned}
\top &:= \bot \rightarrow \bot, \\
\neg\varphi &:= \varphi \rightarrow \bot, \\
\varphi \wedge \psi &:= \neg(\varphi \rightarrow \neg\psi), \\
\varphi \vee \psi &:= \neg\varphi \rightarrow \psi, \\
\varphi \leftrightarrow \psi &:= (\varphi \rightarrow \psi) \wedge (\psi \rightarrow \varphi).
\end{aligned}
$$

In the sequel (D1) will refer to these definitions (here and in the sequel D indicates a definition, R a rule, A an axiom schema and P a propositional axiom schema). In our proof systems we use the following complete axiomatization of propositional logic:

(R1) Modus Ponens: to infer ψ from φ and $\varphi \rightarrow \psi$

(P1) $\varphi \rightarrow (\psi \rightarrow \varphi)$

(P2) $(\varphi \rightarrow (\psi \rightarrow \chi)) \rightarrow ((\varphi \rightarrow \psi) \rightarrow (\varphi \rightarrow \chi))$

(P3) $(\neg\varphi \rightarrow \neg\psi) \rightarrow (\psi \rightarrow \varphi)$.

To this propositional language modal and temporal operators can be added. For modal logic we add two operators: $\mathbf{L}$ (necessarily) and $\mathbf{M}$ (possibly). Temporal logic (in philosophy also known as tense logic) adds four operators: $\mathbf{G}$ (it is always going to be the case), $\mathbf{F}$ (at least once in the future), $\mathbf{H}$ (it has always been the case) and $\mathbf{P}$ (at least once in the past). For a unary operator $\mathbf{O}$ its dual $\overline{\mathbf{O}}$ is defined by

$$
\overline{\mathbf{O}}\,\varphi := \neg\,\mathbf{O}\,\neg\,\varphi.
$$

(Then $\overline{\mathbf{O}_1\,\mathbf{O}_2}$ equals $\overline{\mathbf{O}_1}\,\overline{\mathbf{O}_2}$ and $\overline{\overline{\mathbf{O}}}$ equals $\mathbf{O}$.) The pair $\mathbf{L}, \mathbf{M}$ for modal logic and pairs $\mathbf{G}, \mathbf{F}$ and $\mathbf{H}, \mathbf{P}$ for temporal logic are duals of each other.

The semantics of modal and temporal logic is based on frames and models:

Definition 3.2.1 A frame is a pair (W, R) where W is a non-empty set of 'worlds' and R is a binary relation on W ('alternative' relation or relation of 'accessibility').

A model is a triple (W, R, V) where (W, R) is a frame and V is a valuation on W, that is, it maps proposition letters onto subsets of W (giving the set of worlds where this proposition holds).

Temporal frames are often called point structures $(T, <)$ where T is the set of 'moments' (points in time) and $<$ is the 'precedence' or 'earlier' relation. Usually one imposes at least the restrictions of transitivity and irreflexivity on $<$ giving rise to a strict partial order. Unless stated otherwise we will, however, make no such assumptions and treat $<$ as an arbitrary binary relation. Several notions of validity and semantical consequence are defined as follows.

Definition 3.2.2 A modal formula φ holds in $\mathcal{M} = (W, R, V)$ at $w \in W$, notation $\mathcal{M}, w \models \varphi$, is defined by recursion:

$\mathcal{M}, w \models p$ iff $w \in V(p)$ (for any proposition letter p)
$\mathcal{M}, w \models \bot$ for *no* $\mathcal{M}$ and w
$\mathcal{M}, w \models \varphi \rightarrow \psi$ iff $\mathcal{M}, w \models \varphi \Rightarrow \mathcal{M}, w \models \psi$
$\mathcal{M}, w \models \mathbf{L}\varphi$ iff $\forall w' \in W \, [wRw' \Rightarrow \mathcal{M}, w' \models \varphi]$

For a temporal formula φ, $\mathcal{M} = (T, <, V)$ and $t \in T$ $\mathcal{M}, t \models \varphi$ is defined in the same way except for the replacement of the clause for $\mathbf{L}$ by two clauses for $\mathbf{G}$ and $\mathbf{H}$:

$\mathcal{M}, t \models \mathbf{G}\varphi$ iff $\forall t' \in T \, [t < t' \Rightarrow \mathcal{M}, t' \models \varphi]$
$\mathcal{M}, t \models \mathbf{H}\varphi$ iff $\forall t' \in T \, [t' < t \Rightarrow \mathcal{M}, t' \models \varphi]$.

Furthermore, we define the following derived notions (both for modal and temporal logic):

$\mathcal{M}, w \models \Phi$ if $\forall \varphi \in \Phi \; \mathcal{M}, w \models \varphi$
$\mathcal{M} \models \varphi$ if $\forall w \in W \; \mathcal{M}, w \models \varphi$
$\models \varphi$ if $\forall \mathcal{M} \; \mathcal{M} \models \varphi$ (φ is 'universally valid')
$\mathcal{M} \models \Phi$ if $\forall \varphi \in \Phi \; \mathcal{M} \models \varphi$
$\Phi \models_m \varphi$ if $\forall \mathcal{M} \, [\mathcal{M} \models \Phi \Rightarrow \mathcal{M} \models \varphi]$.

Similar notions can be defined for frames:

$\mathcal{F}, w \models \varphi$ if $\forall V \, (\mathcal{F}, V), w \models \varphi$
$\mathcal{F} \models \varphi$ if $\forall w \in W \; \mathcal{F}, w \models \varphi$
$\mathcal{F} \models \Phi$ if $\forall \varphi \in \Phi \; \mathcal{F} \models \varphi$
$\Phi \models_f \varphi$ if $\forall \mathcal{F} \, [\mathcal{F} \models \Phi \Rightarrow \mathcal{F} \models \varphi]$.

Modal (temporal) formulas express certain constraints on the alternative (precedence) relation in frames where they are valid. When they are interpreted in *models*, modal (temporal) formulas are equivalent to a special kind of formulas in the following first-order language.

Definition 3.2.3 L_1 is the first-order language containing one binary predicate constant R and unary predicate constants $P, P_1, P_2, \ldots, Q, \ldots$.

The binary predicate constant R in this definition corresponds to the alternative relation while the unary predicate constants correspond to the proposition letters $(p, p_1, p_2, \ldots, q, \ldots)$. An example of a modal formula and its first-order equivalent is given by the formula $\mathbf{L}p \to \mathbf{LL}p$ and its L_1-equivalent

$$\forall y(xRy \to Py) \to \forall y(xRy \to \forall z(yRz \to Pz)).$$

The free variable x refers to the current world of evaluation. The general translation τ for modal logic is given in the following definition.

Definition 3.2.4 Let x be a fixed variable.

(i) $\tau(p) = Px$

(ii) $\tau(\neg\varphi) = \neg\tau(\varphi)$

(iii) $\tau(\varphi \wedge \psi) = \tau(\varphi) \wedge \tau(\psi)$

(iv) $\tau(\mathbf{L}\varphi) = \forall y(xRy \to [y/x]\tau(\varphi))$,   where y does not occur in $\tau(\varphi)$.

The translation for temporal logic is similar. τ gives the obvious equivalences (compare the definition of τ with that of $\mathcal{M}, w \models \varphi$ in Definition 3.2.2)

$$\mathcal{M}, w \models \varphi \quad \text{iff} \quad \mathcal{M} \models [w/x]\tau(\varphi)$$
$$\mathcal{M} \models \varphi \quad \text{iff} \quad \mathcal{M} \models \forall x\, \tau(\varphi).$$

The condition in clause (iv) that y should be fresh is not needed. In fact, two variables suffice, for instance the temporal formula $\mathbf{GFH}p$ can be translated into

$$\forall y(x < y \to \exists x(y < x \wedge \forall y(y < x \to Py))).$$

For models we can look for truth-preserving operations, that is, operations on models such that $\mathcal{M}, w \models \varphi$ is preserved. In the sequel, $\mathcal{M}_1 = (W_1, R_1, V_1)$ and $\mathcal{M}_2 = (W_2, R_2, V_2)$.

Definition 3.2.5 $\mathcal{M}_1$ is a *submodel* of $\mathcal{M}_2$ if

(i) $W_1 \subseteq W_2$,

(ii) $R_1 = R_2$ restricted to W_1,

(iii) $V_1(p) = V_2(p) \cap W_1$, for all proposition letters p.

If $\mathcal{M}_1$ has the additional feature that

(iv) W_1 is closed under passing to R_2-successors,

then $\mathcal{M}_1$ is a *generated submodel* of $\mathcal{M}_2$.

The next result is the famous 'Generation Theorem' of [Seg 71].

Theorem 3.2.6 If $\mathcal{M}_1$ is a generated submodel of $\mathcal{M}_2$, then for all $w \in W_1$ and all modal formulas φ:

$$\mathcal{M}_1, w \models \varphi \text{ iff } \mathcal{M}_2, w \models \varphi.$$

The above concerns connections inside one model. For comparing evaluation in different models, we have the following

Definition 3.2.7 A relation Z is a *zigzag connection* between $\mathcal{M}_1$ and $\mathcal{M}_2$ if

(i) domain$(Z) = W_1$, range$(Z) = W_2$,

(ii) if wZv, then w, v verify the same proposition letters,

(iiia) if wZv, and $w' \in W_1$ with wR_1w', then $w'Zv'$ for some $v' \in W_2$ with vR_2v',

(iiib) if wZv, and $v' \in W_2$ with vR_2v', then $w'Zv'$ for some $w' \in W_1$ with wR_1w'.

This notion of zigzag connection is related to the notion of bisimulation (see e.g. [Par 81]). Starting from the basic case (ii), clauses (iiia) and (iiib) ensure that evaluation of modalities in modal formulas yields the same results on either side as is formulated by the following Theorem (see [Seg 70]).

Theorem 3.2.8 If $\mathcal{M}_1$ is zigzag-connected to $\mathcal{M}_2$ by Z, then, for all $w \in W_1, v \in W_2$ with wZv, and all modal formulas φ:

$$\mathcal{M}_1, w \models \varphi \text{ iff } \mathcal{M}_2, v \models \varphi.$$

As we saw above, the standard translation τ for modal logic (see Definition 3.2.4) translates modal formulas into a first-order language L_1. In fact, the translations of modal formulas belong to a smaller class of first-order formulas, called m-formulas, involving restricted quantification. The next result is Theorem 3.9 of [Ben 85]:

Theorem 3.2.9 An L_1-formula α containing at least one free variable is equivalent to an m-formula iff it is invariant for generated submodels and zigzag connections.

Another important technique, the *filtration* method (see e.g. [Seg 71]), relates truth of a formula in a model to truth of that formula in a *finite* model:

Definition 3.2.10 Let $\mathcal{M} = (W, R, V)$ be a model and φ a formula. Ψ is defined to be the finite set consisting of φ together with all its subformulas. For each $w \in W$, set

$$\Psi(w) := \{\psi \in \Psi \mid \mathcal{M}, w \models \psi\}.$$

The *filtrated model* is the model $\mathcal{M}_\Psi = (W_\Psi, R_\Psi, V_\Psi)$ where

$W_\Psi := \{\Psi(w) \mid w \in W\}$ (a finite set),
$\Psi_1 \, R_\Psi \, \Psi_2$ if, for all ψ such that $\mathbf{L}\psi \in \Psi$: $\mathbf{L}\psi \in \Psi_1 \Rightarrow \psi \in \Psi_2$,
$V_\Psi(p) := \{\Psi(w) \mid p \in \Psi(w)\}$.

Theorem 3.2.11 Let $\mathcal{M} = (W, R, V)$ be a model and φ a formula. Define $\Psi, \Psi(w)$ and $\mathcal{M}_\Psi$ as in Definition 3.2.10. Then for all $w \in W$ and all $\psi \in \Psi$:

$$\mathcal{M}, w \models \psi \quad \text{if and only if} \quad \mathcal{M}_\Psi, \Psi(w) \models \psi.$$

The filtration technique can be refined so that R_Ψ preserves certain desirable properties of the original relation R, such as transitivity. Using filtration it is easy to prove the *finite model property*:

Proposition 3.2.12 Any formula which is not universally valid is falsified on some *finite* model.

Proof: Suppose φ is not universally valid. Then there exists some model $\mathcal{M} = (W, R, V)$ and some $w \in W$ such that $\mathcal{M}, w \models \neg\varphi$. Applying filtration to $\mathcal{M}$ and $\neg\varphi$ yields a finite model in which φ is falsified. ∎

For the role of modal (temporal) formulas in expressing constraints on the alternative (precedence) relation in frames, the valuation as given in a particular model is not relevant. To abstract from particular valuations, one simply quantifies universally over the unary predicates in the above translation τ for models. So, when interpreted in *frames*, modal (temporal) formulas get *second-order* transcriptions, with equivalences (say φ contains proposition letters $p_1, \ldots, p_n$):

$$\mathcal{F}, w \models \varphi \quad \text{iff} \quad \mathcal{F} \models \forall P_1 \ldots \forall P_n \, [w/x] \tau(\varphi)$$
$$\mathcal{F} \models \varphi \quad \text{iff} \quad \mathcal{F} \models \forall P_1 \ldots \forall P_n \, \forall x \, \tau(\varphi).$$

For frames we can also look at truth-preserving operations that are related to those for models. We start with the notions of generated subframe and disjoint union. In the sequel, $\mathcal{F}_1 = (W_1, R_1)$ and $\mathcal{F}_2 = (W_2, R_2)$.

Definition 3.2.13 $\mathcal{F}_1$ is a *generated subframe* of $\mathcal{F}_2$ if

(i) $W_1 \subseteq W_2$,

(ii) $R_1 = R_2$ restricted to W_1,

(iii) W_1 is R_2-closed in W_2.

Definition 3.2.14 The *disjoint union* $\oplus \{\mathcal{F}_i \mid i \in I\}$ of a family of frames $\mathcal{F}_i = (W_i, R_i)$ is the disjoint union of the domains W_i, with the obvious coordinate relations R_i.

Notice that each frame $\mathcal{F}_i$ can be viewed as a generated subframe of $\oplus \{\mathcal{F}_i \mid i \in I\}$. The theorem about generated submodels (Theorem 3.2.6 above) now gives the following two results, preservation under generated subframes and preservation under disjoint unions:

Corollary 3.2.15 If $\mathcal{F}_1$ is a generated subframe of $\mathcal{F}_2$, then $\mathcal{F}_2 \models \varphi$ implies $\mathcal{F}_1 \models \varphi$, for all modal formulas φ.

Corollary 3.2.16 If $\mathcal{F}_i \models \varphi$ for all $i \in I$, then $\oplus \{\mathcal{F}_i \mid i \in I\} \models \varphi$, for all modal formulas φ.

The second truth-preserving operation on models concerned zigzag connections. For frames this notion is adapted as follows.

Definition 3.2.17 A *zigzag morphism* from $\mathcal{F}_1$ to $\mathcal{F}_2$ is a function $f : W_1 \to W_2$ satisfying

(i) wR_1w' implies $f(w)R_2f(w')$, that is, f is an ordinary R-homomorphism;

which has the additional property that

(ii) if $f(w)R_2v$, then there exists $u \in W_1$ with wR_1u and $f(u) = v$.

The theorem above about zigzag connections (Theorem 3.2.8) now gives the next result, preservation under zigzag morphisms:

Corollary 3.2.18 If f is a zigzag morphism from $\mathcal{F}_1$ onto $\mathcal{F}_2$, then, for all modal formulas φ, $\mathcal{F}_1 \models \varphi$ implies $\mathcal{F}_2 \models \varphi$.

The last truth-preserving operation on frames is not related to those for models. For it we need the notions of ultrafilter and ultrafilter extension which we define next.

Definition 3.2.19 An *ultrafilter* U on W is a set of subsets of W such that

(i) $X \in U$ or $Y \in U$ if and only if $X \cup Y \in U$,

(ii) $X \notin U$ if and only if $W - X \in U$.

Remark 3.2.20 Ultrafilters are rather unconstructive objects: to prove their existence one needs Zorn's Lemma (or equivalently, the Axiom of Choice). For more information on the esoteric notion of an ultrafilter the reader may consult [CK 73], Chapter 4.

Definition 3.2.21 The *ultrafilter extension* of a frame $\mathcal{F} = (W, R)$, denoted by $ue(\mathcal{F})$, is the frame $(ue(W), ue(R))$ with

(i) $ue(W)$ is the set of all ultrafilters on W,

(ii) $U\ ue(R)\ U'$ if for each $X \subseteq W$ such that $X \in U$,
    $\pi(X) := \{w \in W \mid \exists v \in X\ wRv\} \in U'$.

This leads to the last preservation result: anti-preservation under ultrafilter extensions.

Theorem 3.2.22 If $ue(\mathcal{F}) \models \varphi$, then $\mathcal{F} \models \varphi$, for all modal formulas φ.

For more details on these preservation results, the reader may consult [Ben 84], section 2.1.

The above translation into second-order logic gives rise to two opposite questions: which first-order relational conditions are definable by a modal (temporal) formula, and which modal (temporal) formulas define a first-order relational condition? To be precise, these questions are concerned about modal (temporal) formulas φ and sentences α in the first-order language containing one binary predicate constant R (respectively $<$) and identity $=$ such that

$$\mathcal{F} \models \varphi \quad \text{iff} \quad \mathcal{F} \models \alpha, \quad \text{for all frames } \mathcal{F}.$$

The following is a list of common first-order conditions for the precedence relation (representing assumptions about time):

TRANS: $\forall xyz(x < y < z \rightarrow x < z)$ (transitivity)

IRREF: $\forall x \, \neg \, x < x$ (irreflexivity)

LIN: $\forall xy(x < y \lor x = y \lor y < x)$ (comparability)

L-LIN: $\forall xyz((y < x \land z < x) \rightarrow (y < z \lor y = z \lor z < y))$ (left-linearity)

BEGIN: $\exists x \, \neg \exists y \, y < x$ (a beginning)

END: $\exists x \, \neg \exists y \, x < y$ (an end)

SUC-P: $\forall x \, \exists y \, y < x$ (succession towards past)

SUC-F: $\forall x \, \exists y \, x < y$ (succession towards future)

DENS: $\forall xy(x < y \rightarrow \exists z \, x < z < y)$ (density)

DISC: $\begin{aligned} &\forall xy(x < y \rightarrow \exists z(x < z \land \neg \exists u \, x < u < z)) \\ \land \quad &\forall xy(x < y \rightarrow \exists z(z < y \land \neg \exists u \, z < u < y)) \end{aligned}$ (discreteness).

Of these, the following are definable with temporal logic:

TRANS by $\mathbf{FF}p \rightarrow \mathbf{F}p$,

L-LIN   by $\mathbf{P}p \rightarrow \mathbf{H}(\mathbf{P}p \lor p \lor \mathbf{F}p)$,

SUC-P   by $\mathbf{H}p \rightarrow \mathbf{P}p$,

SUC-F   by $\mathbf{G}p \rightarrow \mathbf{F}p$,

DENS   by $\mathbf{F}p \rightarrow \mathbf{FF}p$.

As an example how one proves such equivalences of first-order relational conditions and temporal formulas we prove here the equivalences for TRANS and DENS.

First suppose that $<$ is transitive and consider any valuation V on $(T, <)$ verifying $\mathbf{FF}p$ in t. By applying the definition of $\mathbf{F}$ twice there exist t' and t'' such that $t < t' < t''$ and t'' verifies p (and t' verifies $\mathbf{F}p$). By transitivity $t < t''$, so $\mathbf{F}p$ is also verified in t. Hence, $\mathbf{FF}p \rightarrow \mathbf{F}p$ holds at arbitrary points for all valuations V.

Conversely, suppose that $\mathbf{FF}p \rightarrow \mathbf{F}p$ holds at t for all valuations V on $(T, <)$. Consider t' and t'' such that $t < t'$ and $t' < t''$. Then, for the particular valuation V assigning precisely $\{t''\}$ to p, $\mathbf{FF}p$ is true at t. Consequently, by the assumption that $\mathbf{FF}p \rightarrow \mathbf{F}p$ is true at t for V it follows that $\mathbf{F}p$ must be true at t for V. This implies the existence of a $t < t'''$ with t''' verifying p. As $V(p)$ consists of t'' only, this means that $t < t''$, so $<$ is transitive.

Next suppose that $<$ is dense and consider any valuation V on $(T, <)$ verifying $\mathbf{F}p$ in t. By the definition of $\mathbf{F}$ there exists t' such that $t < t'$ and p is true at t'. By density, then, there exists some t'' in between: $t < t'' < t'$. So, again by the definition of $\mathbf{F}$, $\mathbf{F}p$ is true at t'' and hence $\mathbf{FF}p$ holds at t. Thus, $\mathbf{F}p \rightarrow \mathbf{FF}p$ holds at arbitrary points for all valuations V.

Conversely, suppose that $\mathbf{F}p \rightarrow \mathbf{FF}p$ holds at t for all valuations V on $(T, <)$. Consider any t' such that $t < t'$. Then, for the particular valuation V assigning precisely $\{t'\}$ to p, $\mathbf{F}p$ is true at t. Consequently, by the assumption that $\mathbf{F}p \rightarrow \mathbf{FF}p$ is true at t for V it follows that $\mathbf{FF}p$ must be true at t for V. This implies the existence of $t < t'' < t'''$ with t''' verifying p. As $V(p)$ consists of t' only, this means that $t < t'' < t'$, so $<$ is dense.

That the others are not definable by a temporal formula can be proved using the above preservation results. To see how such a negative conclusion is reached we prove as an example the cases of IRREF and LIN. For the case of IRREF consider the map from $\mathbb{Z}$ (the set of integers) to the reflexive single element point structure which is a zigzag morphism from an irreflexive point structure onto a reflexive one. For LIN we use the preservation result for disjoint unions: an irreflexive single element point structure is linear, but the disjoint union of two of its copies is not. For further details, see [Ben 83], section II.2.2.

Another interesting topic related to expressive power considerations concerns the possibility to characterize special structures. In this case the expressive power of a language is investigated by examining the ability to discriminate between special structures. For example, modal logic cannot distinguish $\mathbb{Z}$ from $\mathbb{N}$:

Proposition 3.2.23 $(\mathbb{N}, <) \models \varphi \Leftrightarrow (\mathbb{Z}, <) \models \varphi$ for all modal formulas φ.

Proof: If $(\mathbb{Z}, <) \not\models \varphi$, for some modal formula φ, then $(\mathbb{Z}, <, V), w \models \neg\varphi$ for some $w \in \mathbb{Z}$ and some valuation V. The subframe generated by w is isomorphic to $(\mathbb{N}, <)$. So by preservation under generated submodels we have $(\mathbb{N}, <, V'), w \models \neg\varphi$ where $V'(p) = V(p) \cap \mathbb{N}$, for all p.

Conversely, any valuation V on $(\mathbb{N}, <)$ gives rise to a valuation V' on $(\mathbb{Z}, <)$ which is equivalent to V on $(\mathbb{N}, <)$. Therefore, if $(\mathbb{N}, <, V), n \models \neg\varphi$, then $(\mathbb{Z}, <, V'), n \models \neg\varphi$. ■

A related topic is that of categoricity.

Definition 3.2.24 A set Φ of formulas is called *(frame) categorical* if there is, up to isomorphism, only one frame validating Φ. Φ is called λ-*categorical* if, up to isomorphism, Φ has only one frame of power λ validating it.

For modal and temporal logic the notions of categoricity and ω-categoricity are rather meaningless:

Proposition 3.2.25 Let $\mathcal{F} \models \Phi$ where Φ is a set of modal or temporal formulas, and let I be an index set. Then for each $i \in I$, there is a frame $\mathcal{F}_i \models \Phi$ such that $\mathcal{F}_i \not\cong \mathcal{F}_j$ if $i \neq j$.

Proof: Put $\mathcal{F}_0 := \mathcal{F}$. If $i > 0$ and $i \in I$, define $\overline{\lambda_i}$ to be the smallest cardinal λ such that $\lambda > |\mathcal{F}_j|$ holds for all $j < i$. Put $\mathcal{F}_i := \oplus_\kappa \{\mathcal{F} \mid \kappa < \overline{\lambda_i}\}$. A simple counting argument shows that $\mathcal{F}_i \not\cong \mathcal{F}_j$, if $i \neq j$. Furthermore, using the preservation results for modal and temporal logic it can be easily verified that $\mathcal{F}_i \models \Phi$, for each $i \in I$. ■

Proposition 3.2.26 If $\mathcal{T}$ is a modal or temporal theory that is valid on some countably infinite frame, but invalid on every finite frame, then $\mathcal{T}$ is not ω-categorical.

Proof: Let $\mathcal{F}$ be a countably infinite frame with $\mathcal{F} \models \mathcal{T}$. If $\mathcal{F}$ is connected, then $\mathcal{F} \oplus \mathcal{F}$ validates $\mathcal{T}$ without being isomorphic to $\mathcal{F}$. If $\mathcal{F}$ is not connected, then let $w \in W$ and consider the subframe $\mathcal{F}_w$ of $\mathcal{F}$ generated by w. Then $\mathcal{F}_w$ validates $\mathcal{T}$, so by assumption it is countably infinite. Finally, it cannot be isomorphic to $\mathcal{F}$ because it is connected.   ∎

In section 2 of Chapter 4 we will encounter extended modal and temporal logics for which these notions become meaningful.

We now proceed with the second question: which modal (temporal) formulas define a first-order condition? Here, we consider just one aspect of the general issue. Many examples of first-order definable formulas have a common syntactic pattern. A typical instance is the following result from [Sah 75].

Definition 3.2.27 A modal formula is called a *Sahlqvist-form* when it is of the form $\varphi \rightarrow \psi$ where

  (i) φ is constructed from $p, \mathbf{L}p, \mathbf{L}\mathbf{L}p, \ldots, \bot, \top$ using only $\wedge, \vee$ and $\mathbf{M}$, while

 (ii) ψ is constructed from proposition letters, $\bot, \top$ using $\wedge, \vee, \mathbf{M}$ and $\mathbf{L}$.

The basic restrictions imposed by Sahlqvist-forms forbid $\mathbf{L}\mathbf{M}$ or $\mathbf{L}(\ldots \vee \ldots)$ combinations in the antecedent φ.

Theorem 3.2.28 All Sahlqvist-forms define first-order conditions.

Proof: In fact, if χ is a Sahlqvist-form it is *locally equivalent* with a first-order condition α containing precisely one free variable x (as in the standard translation τ for modal logic in Definition 3.2.4), that is to say

$$\forall w \in W \; (\mathcal{F}, w \models \chi \; \text{ iff } \; \mathcal{F}, w \models \alpha), \quad \text{for all frames } \mathcal{F}.$$

Earlier we only introduced *global equivalence* between a modal formula χ and a first-order sentence α defined by

$$\mathcal{F} \models \chi \; \text{ iff } \; \mathcal{F} \models \alpha, \quad \text{for all frames } \mathcal{F}.$$

Local equivalence is stronger than global equivalence: if χ is locally equivalent with α, then χ is globally equivalent with $\forall x \, \alpha$ as one easily checks. For the proof of local equivalence of a Sahlqvist-form with a first-order condition and generalizations thereof we refer the interested reader to [Ben 85], Chapter IX. Here we illustrate the procedure by means of an example. Consider the modal formula

$$\mathbf{L}(\mathbf{L}p \rightarrow q) \vee \mathbf{L}(\mathbf{L}q \rightarrow p)$$

which is an axiom of the modal system $S4.3$. First we have to transform this into an equivalent formula that is a Sahlqvist-form. To that end rewrite the disjunction as an implication of the negation of the first disjunct and the second disjunct:

$$\mathbf{M}(\mathbf{L}p \wedge \neg q) \rightarrow \mathbf{L}(\mathbf{M}\neg q \vee p).$$

Next we have to get rid of the negations. Fortunately the only negations in this formula are the two occurrences of $\neg q$. Therefore we can use a simple lemma stating that for all frames $\mathcal{F}$, worlds $w \in W$ and formulas φ:

$$\mathcal{F}, w \models \varphi \quad \text{iff} \quad \mathcal{F}, w \models [\neg p/p]\varphi \quad \text{for all proposition letters } p.$$

(This lemma follows from a more general substitution lemma: [Ben 85], Lemma 2.5.) Thus we may rewrite our example formula as

$$\mathbf{M}(\mathbf{L}p \wedge q) \rightarrow \mathbf{L}(\mathbf{M}q \vee p).$$

Once we have obtained a formula that is a Sahlqvist-form we apply the standard translation τ for modal formulas of Definition 3.2.4 in such a way that no two quantifiers have the same bound variable. In this case this yields the first-order formula

$$\exists y \, (xRy \wedge \forall z \, (yRz \rightarrow Pz) \wedge Qy) \rightarrow$$
$$\forall s \, (xRs \rightarrow (\exists t \, (sRt \wedge Qt) \vee Ps)).$$

The (outer) existential quantification in the antecedent of this formula is now rewritten as a universal quantification over the whole formula yielding

$$\forall y \, ((xRy \wedge \forall z \, (yRz \rightarrow Pz) \wedge Qy) \rightarrow$$
$$\forall s \, (xRs \rightarrow (\exists t \, (sRt \wedge Qt) \vee Ps))).$$

At this stage the main problem, the presence of the unary predicate constants has to be tackled. This is done by the so-called method of substitutions. To start with, fix a variable not occurring in the first-order formula; in this case we can take u. In the following we concentrate on proposition letters in the antecedent φ of the Sahlqvist-form, in the example $\mathbf{M}(\mathbf{L}p \wedge q)$. Let $\hat{p}$ be an occurrence of p in φ. By $v(\hat{p})$ we denote the bound variable in the standard translation of φ corresponding to the innermost occurrence of $\mathbf{M}$ in φ the scope of which contains $\hat{p}$; or, if no such occurrence of $\mathbf{M}$ exists, $v(\hat{p}) = x$. In our example the only proposition letters in φ, p and q, occur only once and $v(p) = v(q) = y$ since y is the bound variable corresponding to the only occurrence of $\mathbf{M}$ in φ. Now, for the greatest number j such that $\hat{p}$ occurs within a subformula of the form $\mathbf{L}^j p$ (i.e. p prefixed by j times an $\mathbf{L}$), put

$$CV(\hat{p}, \varphi) := v(\hat{p}) \, R^j \, u$$

(where R^0 equals identity $=$). $CV(p, \varphi)$ is defined as the disjunction of all $CV(\hat{p}, \varphi)$, where $\hat{p}$ is an occurrence of p in φ. In our example we obtain

$$CV(p, \varphi): \quad yRu$$
$$CV(q, \varphi): \quad y = u.$$

Finally, the first-order equivalent of the original modal formula is obtained by substituting, for each proposition letter p and corresponding unary predicate constant P, and each individual variable z, $[z/u]CV(p, \varphi)$ for Pz in the obtained standard translation of the original modal formula. So, in our example we substitute respectively in the last first-order formula above:

$$[z/u]CV(p, \varphi) \quad \text{for} \quad Pz,$$
$$[y/u]CV(q, \varphi) \quad \text{for} \quad Qy,$$
$$[t/u]CV(q, \varphi) \quad \text{for} \quad Qt,$$
$$[s/u]CV(p, \varphi) \quad \text{for} \quad Ps.$$

This yields the first-order formula

$$\forall y \, ((xRy \wedge \forall z \, (yRz \rightarrow yRz) \wedge y = y) \rightarrow$$
$$\forall s \, (xRs \rightarrow (\exists t \, (sRt \wedge y = t) \vee yRs))).$$

After simplification this yields the desired first-order equivalent:

$$\forall y \, (xRy \rightarrow \forall s \, (xRs \rightarrow (sRy \vee yRs))). \qquad \blacksquare$$

Remark 3.2.29 Sahlqvist-forms may at first sight seem to be a rather restricted syntactical class of formulas, but as the example in the proof above shows many formulas can be transformed into an equivalent Sahlqvist-form.

We now turn to axiomatizations of modal and temporal logic. In the sequel, by $\Psi \vdash_{PS} \varphi$ we denote that φ is derivable from Ψ within the proof system PS.

Definition 3.2.30 The minimal modal logic proof system K consists of (D1), Modus Ponens (R1), (P1)–(P3) and

(D2) $\mathbf{M}\varphi := \neg\,\mathbf{L}\,\neg\,\varphi$

(A1) $\mathbf{L}(\varphi \rightarrow \psi) \rightarrow (\mathbf{L}\varphi \rightarrow \mathbf{L}\psi)$ (Distribution)

(R2) to infer $\mathbf{L}\varphi$ from φ (Necessitation).

K is called minimal because it precisely axiomatizes $\models_m$:

Theorem 3.2.31 (Completeness of K) For all modal formulas φ and sets of modal formulas Ψ:

$$\Psi \vdash_K \varphi \quad \text{if and only if} \quad \Psi \models_m \varphi.$$

Proof: The Henkin method for proving completeness is well known from the literature and we use techniques from [Ben 83], Theorem II.2.3.6 and [Ben 85], Theorem 6.1 for the proof. A simplified schema of the proof with the main lemmas and propositions is given in Figure 3.1. The easy side

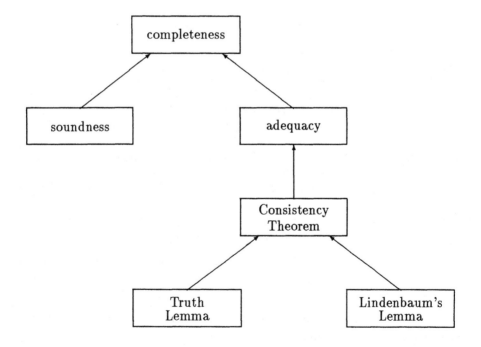

Figure 3.1: Schema for Completeness Proof of K

to a completeness theorem is usually the soundness part: a check whether the axioms and rules were chosen at least correctly. In this case a routine induction establishes that

$$\Psi \vdash_K \varphi \quad \text{implies} \quad \Psi \models_m \varphi.$$

The converse (adequacy) is more exciting. We will prove this by contraposition:

$$\Psi \nvdash_K \varphi \quad \text{implies} \quad \Psi \nvDash_m \varphi,$$

that is, we use the information that φ cannot be derived from Ψ within K to find a model for Ψ in which φ fails at some world. This may be done by reformulating this information more 'positively' through the following notion of consistency.

> A set of formulas Φ is called Ψ-*consistent* if for no finite number
> of formulas $\varphi_1, \ldots, \varphi_n$ from Φ, $\Psi \vdash_K \neg(\varphi_1 \wedge \ldots \wedge \varphi_n)$.

Now, if $\Psi \nvdash_K \varphi$, then $\{\neg\varphi\}$ is obviously Ψ-consistent. So, it suffices to show that Ψ-consistent sets of formulas are satisfiable in at least one model for Ψ, as formulated in the following CONSISTENCY THEOREM:

> For each Ψ-consistent set of formulas Φ, there exists a model
> $\mathcal{M} = (W, R, V)$ for Ψ with some $w \in W$ such that $\mathcal{M}, w \models \Phi$.

Such models may verify more formulas at w than just those in Φ. Therefore we introduce the following additional notion:

> Φ is *maximally Ψ-consistent* if it is Ψ-consistent, while none of
> its proper extensions are.

Notice that maximally Ψ-consistent sets always contain Ψ. The following two observations on Ψ-consistent sets enable us to obtain such sets:

1. if Φ is Ψ-consistent, and φ is any formula, then $\Phi \cup \{\varphi\}$ or $\Phi \cup \{\neg\varphi\}$ is Ψ-consistent,

2. the union of any ascending chain $\Phi_1 \subseteq \Phi_2 \subseteq \ldots$ of Ψ-consistent sets is itself Ψ-consistent.

This leads to the following result, LINDENBAUM'S LEMMA:

> Each Ψ-consistent set of formulas is contained in some maximally
> Ψ-consistent one.

Maximally Ψ-consistent sets exhibit a very useful decomposition behavior with respect to Boolean connectives:

If Φ is maximally Ψ-consistent, then
$$\neg\varphi \in \Phi \qquad \text{if and only if} \quad \varphi \notin \Phi,$$
$$\varphi \rightarrow \psi \in \Phi \qquad \text{if and only if} \quad \varphi \in \Phi \Rightarrow \psi \in \Phi.$$

Now we are ready to define the *Henkin Model* of Ψ, notation HM_Ψ. This canonical model consists of the *Henkin Frame* of Ψ, notation HF_Ψ, together with a valuation V_Ψ which are defined as follows. $HF_\Psi = (W_\Psi, R_\Psi)$ where W_Ψ consists of all maximally Ψ-consistent sets of formulas and R_Ψ is defined by

$$\Phi_1 \, R_\Psi \, \Phi_2 \ \text{if, for all formulas} \ \varphi, \ \ \mathbf{L}\varphi \in \Phi_1 \ \Rightarrow \ \varphi \in \Phi_2.$$

$V_\Psi(p)$ is defined as $\{\Phi \in W_\Psi \mid p \in \Phi\}$.
The definition of the Henkin Model as above is guided by the target equivalence of the TRUTH LEMMA:

For all maximally Ψ-consistent sets Φ, and all formulas φ,

$$HM_\Psi, \Phi \models \varphi \quad \text{if and only if} \quad \varphi \in \Phi.$$

The Truth Lemma, together with Lindenbaum's Lemma, is sufficient to prove the Consistency Theorem:

Let Φ be a Ψ-consistent set of formulas. By Lindenbaum's Lemma Φ can be extended to a maximally Ψ-consistent set Φ_{max}. Since $\Phi \subseteq \Phi_{max}$ the Truth Lemma gives the desired $HM_\Psi, \Phi_{max} \models \Phi$. ($HM_\Psi$ is a model for Ψ since every maximally Ψ-consistent set contains Ψ.)

So it remains to prove the Truth Lemma. This is done by induction on the complexity of φ. The basic step is taken care of by the definition of $V_\Psi(p)$. The steps for the Boolean connectives follow from the above decomposition properties of maximally Ψ-consistent sets. It remains to prove the Truth Lemma for the case that $\varphi \equiv \mathbf{L}\psi$. First, suppose $\mathbf{L}\psi \in \Phi$ and let $\Phi' \in W_\Psi$ such that $\Phi \, R_\Psi \, \Phi'$. Then, by the definition of R_Ψ, $\psi \in \Phi'$. Hence, by the induction hypothesis, $HM_\Psi, \Phi' \models \psi$. Therefore $HM_\Psi, \Phi \models \mathbf{L}\psi$.
Finally, the converse direction.
Suppose $\mathbf{L}\psi \notin \Phi$. A maximally Ψ-consistent set Φ' is to be found such that $\Phi \, R_\Psi \, \Phi'$ and $HM_\Psi, \Phi' \not\models \psi$, that is, by the induction hypothesis, $\psi \notin \Phi'$.

To get such a Φ' it suffices to prove that the set $\{\chi \mid \mathbf{L}\chi \in \Phi\} \cup \{\neg\psi\}$ is
Ψ-consistent. (For in that case, by Lindenbaum's Lemma, there exists some
maximally Ψ-consistent extension of this set which does not contain ψ and
is an R_Ψ-successor of Φ since it contains all χ such that $\mathbf{L}\chi \in \Phi$.)

It is only to prove this statement that the axioms of K come into play in the
following reductio ad absurdum argument:

Let $\mathbf{L}\chi_1, \ldots, \mathbf{L}\chi_k \in \Phi$ and suppose that $\Psi \vdash_K \neg(\chi_1 \wedge \ldots \wedge \chi_k \wedge \neg\psi)$, then
by propositional reasoning $\Psi \vdash_K (\chi_1 \wedge \ldots \wedge \chi_k) \to \psi$.

By Necessitation we get $\Psi \vdash_K \mathbf{L}((\chi_1 \wedge \ldots \wedge \chi_k) \to \psi)$, whence by Distribution
$\Psi \vdash_K (\mathbf{L}\chi_1 \wedge \ldots \wedge \mathbf{L}\chi_k) \to \mathbf{L}\psi$.

Since Φ is maximally Ψ-consistent, all formulas derivable from Ψ within K
belong to Φ, in particular the latter formula. By applying Modus Ponens we
conclude that $\mathbf{L}\psi \in \Phi$, a contradiction. ∎

Definition 3.2.32 The minimal temporal logic proof system K_t consists of
(D1), Modus Ponens (R1), (P1)–(P3) and

(D2) $\mathbf{F}\varphi := \neg\mathbf{G}\neg\varphi$
(D3) $\mathbf{P}\varphi := \neg\mathbf{H}\neg\varphi$

(A1) $\mathbf{G}(\varphi \to \psi) \to (\mathbf{G}\varphi \to \mathbf{G}\psi)$
(A2) $\mathbf{H}(\varphi \to \psi) \to (\mathbf{H}\varphi \to \mathbf{H}\psi)$ (distribution of tense)

(A3) $\varphi \to \mathbf{GP}\varphi$
(A4) $\varphi \to \mathbf{HF}\varphi$ (tense mixing)

(R2) to infer $\mathbf{G}\varphi$ from φ
(R3) to infer $\mathbf{H}\varphi$ from φ (temporalization)

Again, K_t is called minimal because it precisely axiomatizes $\models_m$:

Theorem 3.2.33 (Completeness of K_t) For all temporal formulas φ and
sets of temporal formulas Ψ:

$$\Psi \vdash_{K_t} \varphi \quad \text{if and only if} \quad \Psi \models_m \varphi.$$

Proof: The proof is very similar to that for the system K above, the main
difference being the replacement of the definition of R_Ψ by $<_\Psi$ as follows:

$\Phi_1 <_\Psi \Phi_2$ if, for all formulas φ,
$$G\varphi \in \Phi_1 \Rightarrow \varphi \in \Phi_2 \quad \text{and} \quad H\varphi \in \Phi_2 \Rightarrow \varphi \in \Phi_1. \quad \blacksquare$$

The above effective axiomatizations together with the finite model property (established via filtration, see Proposition 3.2.12) guarantee that these logics are decidable: for each formula φ, by simultaneously enumerating all theorems (using the axiomatization) and all finite models, we can check whether φ is a theorem or not (since in the latter case φ is falsified on some finite model by the finite model property). For more details on methods of proving decidability, see, for example, [Bur 84] section 3.

For obtaining similar results about complete axiomatizations of $\models_f$ (instead of $\models_m$) we add the following rule of substitution to our proof systems (for an explanation, see [Ben 83], section II.2.3):

(R4) to infer $[\psi_1/p_1, \ldots, \psi_n/p_n]\,\varphi$ from φ.

We call Φ *frame-complete* if

$$\Phi \vdash^s \varphi \text{ iff } \Phi \models_f \varphi \quad \text{for all } \varphi,$$

where $\vdash^s$ stands for one of the above proof systems with the additional rule of substitution. For $\models_m$ we obtained general completeness results, that is for all Φ. For $\models_f$, however, this is not possible: there are Φ which are not frame-complete. So, apart from the general completeness theorems above, modal (temporal) completeness theorems are dealing with special Φ that are frame-complete. Many such results involve a slight generalization of frame-completeness, namely completeness with respect to a class of frames which is defined as follows.

Definition 3.2.34 Let $\mathcal{C}$ be a class of frames. Φ is called complete with respect to $\mathcal{C}$ if

$$\Phi \vdash^s \varphi \text{ iff } \forall \mathcal{F}(\mathcal{F} \in \mathcal{C} \Rightarrow \mathcal{F} \models \varphi) \quad \text{for all } \varphi.$$

Note that Φ frame-complete reduces to Φ complete with respect to $\{\mathcal{F} \mid \mathcal{F} \models \Phi\}$.

3.3 Temporal Logic with **until** and **since**

We first define the syntax of propositional temporal logic with **until** and **since** operators.

Definition 3.3.1 $L(\mathbf{until}, \mathbf{since})$ is the language with

vocabulary: atomic propositions $P_0, P_1, \ldots$
                logical operators $\neg$, $\wedge$, $\mathbf{until}$, $\mathbf{since}$

formulas: $P_i (i \in \mathbb{N})$
          $\neg\varphi_1$, $\varphi_1 \wedge \varphi_2$, $\varphi_1 \mathbf{\ until\ } \varphi_2$ and $\varphi_1 \mathbf{\ since\ } \varphi_2$ (φ_1, φ_2 formulas).

To give the semantics of $L(\mathbf{until}, \mathbf{since})$ we can use the notions of frames, valuations and models introduced in section 2 (see Definition 3.2.1). For languages with $\mathbf{until}$ and $\mathbf{since}$ we will suppose the temporal frames to be strict partial orders, that is, $<$ is transitive and irreflexive. In the definition of $\mathcal{M}, t \models \varphi$ we have only to include the following clauses for the operators $\mathbf{until}$ and $\mathbf{since}$:

$$\mathcal{M}, t \models \varphi_1 \mathbf{\ until\ } \varphi_2 \quad \text{iff} \quad \exists\, t' \in T[t < t' \text{ and } \mathcal{M}, t' \models \varphi_2 \text{ and}$$
$$\forall\, t'' \in T[t < t'' < t' \Rightarrow \mathcal{M}, t'' \models \varphi_1]]$$
$$\mathcal{M}, t \models \varphi_1 \mathbf{\ since\ } \varphi_2 \quad \text{iff} \quad \exists\, t' \in T[t' < t \text{ and } \mathcal{M}, t' \models \varphi_2 \text{ and}$$
$$\forall\, t'' \in T[t' < t'' < t \Rightarrow \mathcal{M}, t'' \models \varphi_1]].$$

Because of irreflexivity of $<$ the operators $\mathbf{until}$ and $\mathbf{since}$ will also be irreflexive, that is, they do not include the present as part of the future.

Kamp (see [Kam 68]) proves that $L(\mathbf{until}, \mathbf{since})$ is expressively complete with respect to the class of complete linear orders. For the class of ω-models (obtained by allowing as the only temporal frame the natural numbers with their usual ordering) it is shown in [GPSS 80] that only $\mathbf{until}$ as temporal operator already suffices for expressive completeness. The temporal operators $\mathbf{F}, \mathbf{G}, \mathbf{P}, \mathbf{H}$ of section 2 can be defined easily in terms of $\mathbf{until}$ and $\mathbf{since}$:

$$\mathbf{F}\,\varphi \ := \ \top \mathbf{\ until\ } \varphi$$

$$\mathbf{P}\,\varphi \ := \ \top \mathbf{\ since\ } \varphi$$

where still $\mathbf{G} \equiv \overline{\mathbf{F}}$ and $\mathbf{H} \equiv \overline{\mathbf{P}}$, of course.

In an unpublished paper ([Sta 79]) Stavi introduced two additional operators $\widehat{\mathbf{until}}$ and $\widehat{\mathbf{since}}$ in order to improve the above expressive completeness result of Kamp to the class of all linear orders. These operators are defined by

$$\mathcal{M}, t \models \varphi_1 \widehat{\mathbf{\ until\ }} \varphi_2 \quad \text{iff}$$
$$\exists\, t''[t < t'' \text{ and } \forall\, t'[t < t' < t'' \Rightarrow \mathcal{M}, t' \models \varphi_1]] \text{ and}$$

$$\forall t''[(t < t'' \text{ and } \forall t'[t < t' < t'' \Rightarrow \mathcal{M}, t' \models \varphi_1]) \Rightarrow$$
$$(\mathcal{M}, t'' \models \varphi_1 \text{ and}$$
$$\exists t''''[t'' < t'''' \text{ and } \forall t'''[t'' < t''' < t'''' \Rightarrow \mathcal{M}, t''' \models \varphi_1]])] \text{ and}$$
$$\exists t'''[t < t''' \text{ and } \mathcal{M}, t''' \models \neg\varphi_1 \text{ and } \mathcal{M}, t''' \models \varphi_2 \text{ and}$$
$$\forall t''[(t < t'' < t''' \text{ and } \exists t'[t < t' < t'' \text{ and } \mathcal{M}, t' \models \neg\varphi_1])$$
$$\Rightarrow \mathcal{M}, t'' \models \varphi_2]]$$

$$\mathcal{M}, t \models \varphi_1 \widehat{\text{ since }} \varphi_2 \quad \text{iff}$$
$$\exists t''[t'' < t \text{ and } \forall t'[t'' < t' < t \Rightarrow \mathcal{M}, t' \models \varphi_1]] \text{ and}$$
$$\forall t''[(t'' < t \text{ and } \forall t'[t'' < t' < t \Rightarrow \mathcal{M}, t' \models \varphi_1]) \Rightarrow$$
$$(\mathcal{M}, t'' \models \varphi_1 \text{ and}$$
$$\exists t''''[t'''' < t'' \text{ and } \forall t'''[t'''' < t''' < t'' \Rightarrow \mathcal{M}, t''' \models \varphi_1]])] \text{ and}$$
$$\exists t'''[t''' < t \text{ and } \mathcal{M}, t''' \models \neg\varphi_1 \text{ and } \mathcal{M}, t''' \models \varphi_2 \text{ and}$$
$$\forall t''[(t''' < t'' < t \text{ and } \exists t'[t'' < t' < t \text{ and } \mathcal{M}, t' \models \neg\varphi_1])$$
$$\Rightarrow \mathcal{M}, t'' \models \varphi_2]].$$

Intuitively, **until** and **since** take care of closing the 'gaps' in incomplete linear orders. An informal explanation of the **until** operator (and similarly for **since**) is the following. $\varphi_1 \widehat{\text{ until }} \varphi_2$ asserts the existence of a 'gap' ahead (i.e. in the future) in the ordering such that

1. from the current moment up till that gap φ_1 will be true (this follows from the first two conjuncts of the definition),

2. the gap is approached from the right (i.e. from the future) both by $\neg\varphi_1$ and by φ_2, that is to say no matter how near we take a point after the gap, there will be a point where $\neg\varphi_1$ (and the same for φ_2) holds in between that point and the gap (the part about φ_2 follows from the third conjunct of the definition, while the part about $\neg\varphi_1$ and the existence of the gap follow from all three conjuncts of the definition).

Similarly to the extension of propositional logic with the temporal operators **until** and **since** one can extend predicate logic with these operators to get a first-order temporal logic. In first-order temporal logics problems arise because of the interplay between quantification and time (see e.g. [Gar 84],[Coc 84]). One of these problems is the possibility that the quantified variables (and possibly even their value domains) change over time. We avoid this problem by only allowing quantification over variables that do not change over time (often called 'global' variables in contrast with 'local' variables). Even in this restricted case most first-order temporal logics are

incomplete (usually shown by proving that Peano Arithmetic can be encoded
into them).

3.4   Temporal Logic in Computer Science

This section is not intended as a brief overview but serves as a background for
the motivation of certain decisions to adapt temporal logic in later chapters.

Since the seminal paper of Pnueli ([Pnu 77]) the use of temporal logic for
reasoning about many types of computerized systems and programs has been
steadily increasing. This can be explained by the fact that the underlying
semantics of temporal logic fits well with the notion of computation as used
in computer science as we will show now. Temporal logic is intended for
reasoning about situations changing in time. Its semantics makes a clear distinction between the static aspect of a situation, represented by a state, and
the dynamic aspect, the relation (in time) between states. This distinction
is also reflected in the syntax: a state is described by the classical part of
temporal logic while the temporal operators are used for the description of
the evolution of the situation over time. In this way states and time need not
be introduced explicitly in the logic itself. The connection with the notion of
computation is that a computation can be seen as a sequence of states where
each transition from one state to the next state in the sequence (each step
of the computation) can be thought of as a tick of some computation clock.
In this view computer systems are described as generators of computations
(also called execution sequences). Therefore, the applications of temporal
logic in computer science are usually restricted to the class of discrete systems where an execution of a system can be viewed indeed as a sequence of
state transitions. For that reason the temporal frames considered are also
discrete.

The two most common types of temporal frames used in computer science
are the natural numbers with their usual ordering and tree-like structures
where branching is allowed only towards the future, giving rise, respectively,
to what is commonly called linear (time) temporal logic and branching time
temporal logic. Concerning the list of common first-order conditions on the
precedence relation representing assumptions about time in section 2, we
see then that the temporal frames of linear temporal logic obey TRANS,
IRREF, LIN and DISC (and usually also BEGIN and SUC-F) while those
of branching time temporal logic obey TRANS, IRREF, L-LIN and DISC

(and again usually also BEGIN and SUC-F). For a comparison between linear and branching time temporal logic, see, for example, [Sti 87]. Apart from linear and branching time temporal logic there are temporal logics in use in computer science that are based on other types of temporal frames, for example, the partial order temporal logic of [PW 84], the temporal logic for event structures of [Pen 88] and the interleaving set temporal logic (using a mixture of branching time and partial order elements) of [KP 87], but these form only a minority. Another approach is the one where temporal frames are not based on points but on intervals instead. This approach is also represented in computer science (for an excellent overview of the interval-based approach vs. the point-based approach in philosophy see [Ben 83]), for example, Interval Temporal Logic with its executable subset Tempura of Moszkowski (see [Mos 83],[MM 84],[Mos 86]) and the interval logic of Schwartz et al. (see [SMV 83]).

In this monograph we restrict our attention to temporal logics based on temporal frames with a precedence relation that is linear, in other words we look only at linear time-like temporal logics. [Pnu 77] contains some deviations from classical temporal logic (as treated in section 2), in particular:

1. the present is considered as part of the future and correspondingly the basic temporal operators are reflexive,

2. only future temporal operators are used.

In the sequel we will denote the reflexive counterparts of the temporal operators $\mathbf{F}$ and $\mathbf{G}$ of classical temporal logic as treated in section 2 by their usual representation in computer science $\Diamond$, respectively $\Box$. In general, irreflexive temporal operators have more expressive power than the corresponding reflexive ones (in section 2 of Chapter 4 the reflexive closure of general modal/temporal operators will be given). Although not done in [Pnu 77] several later papers have included the operators $\mathbf{X}$ (next) and $\mathbf{Y}$ (previous) for indicating the next, respectively previous, element in the precedence relation (remember that in this section this relation is supposed to be a discrete ordering). Over the natural numbers the irreflexive operators can then be expressed, for instance $\mathbf{F}\varphi \equiv \mathbf{X}\Diamond\varphi$. The operators $\mathbf{X}$ and $\mathbf{Y}$ also have their deficiencies, however. For example, these operators lack the abstractness needed to achieve a fully abstract semantics (i.e., on a level of abstraction equalling that of a semantics formalizing the chosen notion of observable behavior) of concurrent programs (see [Lam 83a],[BKP 86]).

Concerning the second deviation above, it can be shown (see the results about expressive completeness in section 3) that the temporal operator **until** already suffices for expressive completeness over the natural numbers. Therefore, from the viewpoint of expressive power there is no need to introduce past operators when working over the natural numbers. However, [KVR 83] showed the advantages of such operators for the elegant specification of message passing systems (see section 6 of Chapter 5) and [LPZ 85] contains many theoretical results about the usefulness of past operators.

We now come back on the topic of temporal logic as a specification language for computerized systems and programs. As we have seen above, a computation of a computer system can be described as a (linear) sequence of states and associated events (state transitions). In linear temporal logic the approach is taken that the behavior of a system S is given by the set of its computations, say Σ. A temporal formula φ is then defined to be valid for S (φ is a valid property of S) if each $\sigma \in \Sigma$ satisfies φ, that is, $\sigma \models \varphi$ in the sense of Definition 3.2.2 in section 2 (remember that the underlying time domain of linear temporal logic is the set of natural numbers so that the sequence σ can function as a model). Recently, a slightly different approach, the anchored version of temporal logic, has been advocated in [MP 89]. In this approach a temporal formula φ is defined to be valid for a system S with set of computations Σ if φ holds in σ at the initial point for each $\sigma \in \Sigma$, that is, $\sigma, 0 \models \varphi$ for all $\sigma \in \Sigma$ (again in the sense of Definition 3.2.2 in section 2). The difference between the two versions of temporal validity is that the first requires a temporal formula to hold at *all* points in all computations while the latter requires this only for the *first* point of all computations. This difference is largely a matter of taste because the two versions can be reduced to each other, for example from not anchored to anchored:

$$\sigma \models \varphi \quad \Leftrightarrow \quad \sigma, 0 \models \Box \varphi.$$

Although the anchored version of temporal logic is often more convenient to use in practice, we do not adopt it in this monograph because formulas in our specification examples are intended to hold at all points in all computations (thereby avoiding the need to prefix all formulas by the $\Box$-operator which would be required in the anchored version).

In Chapter 2 data elements were partitioned into two categories, namely state variables and events. For the description of data elements we need a first-order variant of linear temporal logic. This variant partitions the set of variables into so-called global and local variables where quantification is only allowed over global variables (so the local variables always occur as free variables). Global variables range over fixed data domains and serve to denote elements thereof while local variables model the state variables (such as variables occurring in a program). Events are modeled as predicates (where the parameters of the event become the arguments of the predicate).

When using temporal logic for the specification of programs, a fundamental classification of program properties differentiates between safety- and liveness-properties. For a syntactical classification of temporal properties into a hierarchy refining this safety-liveness classification, see [MP 87]. Characterizations and decidability of safety- and liveness-properties using connections with model theory, formal language theory and semigroup theory are contained in [Tho 86].

To end our account of the application of temporal logic as a specification language in computer science, we can test temporal logic against the requirements for a general specification language in Chapter 2. Syntactical abstractness can be achieved by restricting the local variables and predicates to the state variables and events, respectively, of the interface. Sections 2 and 3 of this chapter witness the formality of temporal logic. Furthermore, temporal logic is clearly a uniform formalism. The conformity requirement is illustrated in section 5 of Chapter 5 for message passing systems. Temporal logic is a simple and elegant extension of propositional logic (predicate logic in case of first-order temporal logic), yet powerful enough to express interesting properties of programs such as safety- and liveness-properties. At last, papers such as [Lam 83b],[BKP 84],[BK 85a],[BK 85b] show that temporal logic can be used for hierarchical development in a compositional and modular style.

Chapter 4

Polymodal Logics with Inequality

4.1 Introduction

As has been demonstrated in the previous chapter, modal and temporal logic cannot define all the natural assumptions one would like to make on the alternative and precedence relation, respectively. This state of affairs provides the motivation for this chapter.

The semantics of modal and temporal logic is based on one binary relation, the alternative, respectively, precedence relation. A straightforward generalization of this is to allow several binary relations and corresponding operators leading to polymodal logics (cf. dynamic logic, see [Har 84]). In fact, temporal logic can be viewed as a bimodal logic with precedence relation $<$ and its converse $>$. This chapter studies polymodal logics including the special relation of inequality (this immediately includes also the total relation on worlds). We adapt several results from modal and temporal logic to polymodal models and frames: we provide translations to classical logic (first-order for models, second-order for frames) and adapt the usual zigzag-relation for models to a kind of enriched bisimulation. It turns out that most of the previous preservation results for frames become invalid. This indicates that the addition of operators for inequality considerably increases the expressive power of modal and temporal logic, a fact that is substantiated by showing how several first-order conditions that were not definable in modal (temporal) logic become definable when operators for inequality are added. In fact, all *universal* first-order conditions and several more complicated properties become definable. However, we can still not define all first-order properties: by filtration we can show that the existence of a reflexive world cannot be defined. Apart from conditions on the alternative (precedence) relation, the addition of operators for inequality also allows to express all finite cardinalities of the set of worlds (moments). Conversely, for frames we transfer an existing syntactic characterization for modal (temporal) formulas that can be defined by a first-order property to polymodal formulas.

For three types of polymodal models (with only the inequality relation, with both the alternative and inequality relation, and with both the precedence and inequality relation) we provide complete axiomatizations and show that the resulting polymodal logics are decidable. Also for three special classes of models (linear orderings, dense linear orderings without endpoints and models isomorphic to the integers) completeness results are presented. At last we look at completeness results for classes of frames.

The rest of this chapter is structured as follows. Section 2 introduces our polymodal logics including inequality and investigates semantic issues. In section 3 we look at complete axiomatizations for these logics and show their decidability. Section 4 contains some conclusions.

4.2   Semantics

For the semantic definition of operators in polymodal logics we make the dependence on a particular binary relation R explicit in the following way. For ease of presentation we assume a model $\mathcal{M} = (W, R, V)$ to be fixed. First we abbreviate $\mathcal{M}, w \models \varphi$ by $\varphi(w)$. Relative to R the necessity and possibility operators are defined respectively by

$$\mathbf{L}^R \varphi(w) := \forall w' \in W \, [wRw' \Rightarrow \varphi(w')]$$
$$\text{and}$$
$$\mathbf{M}^R \varphi(w) := \exists w' \in W \, [wRw' \text{ and } \varphi(w')].$$

For temporal logic with operators $\mathbf{F}$, $\mathbf{P}$ (and duals $\mathbf{G}$,$\mathbf{H}$) we have $\mathbf{F} \equiv \mathbf{M}^<$ and $\mathbf{P} \equiv \mathbf{M}^>$. By $\mathrm{PML}(R_1, \ldots, R_n)$ we denote the polymodal logic with operators $\mathbf{M}^{R_1}, \ldots, \mathbf{M}^{R_n}$ and their duals $\mathbf{L}^{R_1}, \ldots, \mathbf{L}^{R_n}$.

For the polymodal operators $\mathbf{M}^R$ and $\mathbf{L}^R$ there is a standard way to make these reflexive by reflexive closure as follows:

$$\dot{\mathbf{M}}^R \varphi := \varphi \lor \mathbf{M}^R \varphi$$
$$\text{and}$$
$$\dot{\mathbf{L}}^R \varphi := \varphi \land \mathbf{L}^R \varphi.$$

The nomenclature stems from the observation that $\dot{\mathbf{M}}^R \equiv \mathbf{M}^{\dot{R}}$ and $\dot{\mathbf{L}}^R \equiv \mathbf{L}^{\dot{R}}$ where $\dot{R}$ is the reflexive closure of the relation R:

$$w \dot{R} w' \quad \text{iff} \quad w = w' \text{ or } w R w'.$$

In this section we consider the special binary relation of inequality, first as the only binary relation and next as an additional relation besides the alternative and precedence relation of modal, respectively temporal logic. Syntactically we define a new operator $\mathbf{D}$ (at a *different* world/moment) corresponding to $\mathbf{M}^{\neq}$:

$$\mathbf{D}\,\varphi(w) \ := \ \exists w' \in W \,[w \neq w' \text{ and } \varphi(w')].$$

Its dual $\overline{\mathbf{D}}$ corresponds of course to $\mathbf{L}^{\neq}$. From $\mathbf{D}$ two very useful operators, $\mathbf{E}$ (there *exists* a world/moment) and its dual $\mathbf{A}$ (for *all* worlds/moments) are defined by reflexive closure:

$$\mathbf{E}\,\varphi \ := \ \varphi \vee \mathbf{D}\,\varphi$$
$$\text{and}$$
$$\mathbf{A}\,\varphi \ := \ \varphi \wedge \overline{\mathbf{D}}\,\varphi.$$

Note that the semantics of $\mathbf{E}\,\varphi$ and $\mathbf{A}\,\varphi$ is independent of the world in which it is evaluated. In fact, $\mathbf{E} \equiv \mathbf{M}^{W \times W}$ and $\mathbf{A} \equiv \mathbf{L}^{W \times W}$. The following table summarizes the operators $\mathbf{L}^R$ and $\mathbf{M}^R$ for four special choices of R.

R	$\mathbf{L}^R\,\varphi$	$\mathbf{M}^R\,\varphi$
$\emptyset$	$\top$	$\perp$
$=$	φ	φ
$\neq$	$\overline{\mathbf{D}}\,\varphi$	$\mathbf{D}\,\varphi$
$W \times W$	$\mathbf{A}\,\varphi$	$\mathbf{E}\,\varphi$

The last two rows cannot be represented without the $\mathbf{D}$-operator. Notice that $=$ and $W \times W$ are the reflexive closure of $\emptyset$ and $\neq$, respectively.

When modal (temporal) formulas are interpreted in *models*, they are equivalent to a special kind of first-order formulas (see section 2 of Chapter 3). Adding operators for inequality does not change this picture. We can simply add a clause in the translation τ (see Definition 3.2.4) for the $\mathbf{D}$-operator:

$$\tau(\mathbf{D}\varphi) \ := \ \exists y(x \neq y \wedge [y/x]\tau(\varphi)),$$

where y does not occur in $\tau(\varphi)$. But again, two variables actually suffice, for example, **GDH**p can be translated into

$$\forall y(x < y \rightarrow \exists x(y \neq x \land \forall y(y < x \rightarrow Py))).$$

Again τ gives the equivalences

$$\mathcal{M}, w \models \varphi \quad \text{iff } \mathcal{M} \models [w/x]\tau(\varphi)$$
$$\mathcal{M} \models \varphi \quad \text{iff } \mathcal{M} \models \forall x\, \tau(\varphi).$$

A semantical characterization on models can be obtained by giving relations between models such that the special first-order formulas from the translation τ above are invariant under these relations. In section 2 of Chapter 3 we defined (Definition 3.2.5) the truth-preserving operations of generated submodels. Clearly these operations can no longer be truth-preserving in the presence of inequality. However, the other truth-preserving operations, zigzag connections, can be adapted as follows.

Definition 4.2.1 A relation Z is an *extended zigzag connection* between $\mathcal{M}_1 = (W_1, R_1, V_1)$ and $\mathcal{M}_2 = (W_2, R_2, V_2)$ if

(i) domain$(Z) = W_1$, range$(Z) = W_2$,

(ii) if wZv, then w, v verify the same proposition letters,

(iiia) if wZv, and $w' \in W_1$ with wR_1w', then $w'Zv'$ for some $v' \in W_2$ with vR_2v',

(iiib) if wZv, and $v' \in W_2$ with vR_2v', then $w'Zv'$ for some $w' \in W_1$ with wR_1w'.

(iva) if wZv, and $w' \in W_1$ with $w \neq w'$, then $w'Zv'$ for some $v' \in W_2$ with $v \neq v'$,

(ivb) if wZv, and $v' \in W_2$ with $v \neq v'$, then $w'Zv'$ for some $w' \in W_1$ with $w \neq w'$.

The only difference with Definition 3.2.7 consists of the additional clauses (iva) and (ivb). These additional clauses impose a strong connection between $\mathcal{M}_1$ and $\mathcal{M}_2$:

(1) if $Z \neq \emptyset$ then domain$(Z) = W_1$, range$(Z) = W_2$,

(2) if wZv, then either this is the only Z-connection for w and v, or both w and v have at least two Z-mates.

So, if Z is non-empty it may be split up in one bijective part where $w \in W_1$ has only one Z-related $v \in W_2$ (and vice versa) and several clusters of Z-related worlds such that each world in such a cluster is Z-related to at least two worlds (of the other model) in that cluster. On top of this one still has conditions (i)–(iii) so that for instance Z-related worlds must verify the same proposition letters. This adaptation of the notion of zigzag connection leads to a corresponding adaptation of Theorem 3.2.8:

Theorem 4.2.2 If Z is an extended zigzag connection between $\mathcal{M}_1$ and $\mathcal{M}_2$, then, for all $w \in W_1$, $v \in W_2$ with wZv, and all formulas φ from $\mathrm{PML}(R, \neq)$:

$$\mathcal{M}_1, w \models \varphi \quad \text{iff} \quad \mathcal{M}_2, v \models \varphi.$$

Section 2 of Chapter 3 contains a theorem (Theorem 3.2.9) stating that the translations of modal formulas into first-order formulas are characterized by their invariance under generated submodels and zigzag connections. Since generated submodels are not truth-preserving operations any more, the natural question for the addition of inequality is whether translations of formulas of $\mathrm{PML}(R, \neq)$ (and similarly for $\mathrm{PML}(<, >, \neq)$) are characterized by their invariance under the above extended zigzag connections. In the meantime a positive answer to this question has been given in [Rijk 89]. Further results on the definability of classes of frames and classes of models are contained in section 4 of [Rijk 90].

When modal formulas are interpreted in *frames*, they become second-order formulas (say φ contains the proposition letters $p_1, \ldots, p_n$):

$$\mathcal{F} \models \varphi \quad \text{iff} \quad \mathcal{F} \models \forall P_1 \ldots \forall P_n \, \forall x \, \tau(\varphi).$$

For frames we can look at preservation results (see section 2 of Chapter 3) such as preservation under disjoint unions:

$$\text{If } \mathcal{F}_i \models \varphi \text{ for all } i \in I, \text{ then } \oplus\{\mathcal{F}_i \mid i \in I\} \models \varphi, \text{ for all } \varphi.$$

Adding inequality destroys most of the previous preservation results: no preservation under generated subframes, nor under disjoint unions, nor under zigzag morphisms. For example, consider the single-world frame $\mathcal{F} = (\{w\}, R)$. Then $\mathcal{F} \models \neg \mathbf{D} \top$ but the disjoint union of two copies of $\mathcal{F}$ does not. This is an indication that adding operators for inequality considerably increases the expressive power of modal and temporal logic. However, anti-preservation under ultrafilter extensions is preserved as was proven in the meantime by Maarten de Rijke ([Ben 89],[Rijk 89]).

The next two questions concern the correspondence over frames between modal formulas and first-order formulas: which modal formulas are defined by a first-order formula and which first-order formulas can be defined by a modal formula? To start with the latter question, section 2 of Chapter 3 lists several common first-order properties of the precedence relation and states which of them are definable with temporal logic. We now show that the addition of operators for inequality allows also the remaining first-order conditions to be defined:

IRREF: $\mathbf{F}p \rightarrow \mathbf{D}p$ (irreflexivity)

LIN: $\mathbf{D}p \rightarrow (\mathbf{P}p \vee \mathbf{F}p)$ (comparability)

BEGIN: $\mathbf{E}\,\mathbf{H} \perp$ and   END: $\mathbf{E}\,\mathbf{G} \perp$ (a beginning and an end)

DISC: $\begin{aligned}&(\mathbf{P}(p \wedge \neg \mathbf{D}p) \rightarrow \mathbf{E}(\mathbf{P}p \wedge \neg \mathbf{P}\mathbf{P}p)) \\ \wedge\;&(\mathbf{F}(p \wedge \neg \mathbf{D}p) \rightarrow \mathbf{E}(\mathbf{F}p \wedge \neg \mathbf{F}\mathbf{F}p))\end{aligned}$ (discreteness).

As examples we prove the equivalences for IRREF and BEGIN.

.First suppose that $<$ is irreflexive and consider any valuation V on $(T, <)$ verifying $\mathbf{F}p$ in t. By the definition of $\mathbf{F}$ there exists t' such that $t < t'$ and p is true at t'. By irreflexivity $t' \neq t$, so by the definition of $\mathbf{D}$, $\mathbf{D}p$ is true at t. Thus, $\mathbf{F}p \rightarrow \mathbf{D}p$ holds at arbitrary points for all valuations V.

Conversely, suppose that $\mathbf{F}p \rightarrow \mathbf{D}p$ holds at t for all valuations V on $(T, <)$. Consider any t' such that $t < t'$. Then, for the particular valuation V assigning precisely $\{t'\}$ to p, $\mathbf{F}p$ is true at t. Consequently, by the assumption that $\mathbf{F}p \rightarrow \mathbf{D}p$ is true at t for V it follows that $\mathbf{D}p$ must be true at t for V. This implies the existence of $t'' \neq t$ with t'' verifying p. As $V(p)$ consists of t' only, this means that $t' \neq t$, so $<$ is irreflexive.

Next suppose that $<$ has a beginning, say t_0. Then for all t it is the case that $t < t_0$ is false. By the definition of $\mathbf{H}$ it follows that $\mathbf{H} \perp$ is true at t_0. Thus, by the definition of $\mathbf{E}$, $\mathbf{E}\,\mathbf{H} \perp$ holds at arbitrary points for all valuations V.

Conversely, suppose that $\mathbf{E}\,\mathbf{H}\,\bot$ is true at t. By the definition of $\mathbf{E}$ there exists a point, say t_0, such that $\mathbf{H}\bot$ holds at t_0. By the definition of $\mathbf{H}$ this implies that there exists no t such that $t < t_0$ since such a t would have to verify $\bot$. Thus, t_0 is a beginning of $<$.

The opposites of BEGIN and END (succession towards past, respectively future) are already definable in temporal logic (see section 2 of Chapter 3) but can now be defined as opposites of BEGIN and END indeed:

$$\text{SUC-P: } \mathbf{A}\,\mathbf{P}\,\top$$
$$\text{and}$$
$$\text{SUC-F: } \mathbf{A}\,\mathbf{F}\,\top.$$

Irreflexivity and comparability are examples of universal first-order conditions. In fact:

Theorem 4.2.3 All universal first-order conditions on $R, =$ are definable in $\text{PML}(R, \neq)$.

Proof: For $\forall x_1 \ldots \forall x_n\ BOOL(x_i R x_j, x_i = x_j)$ take

$$(\mathbf{U}p_1 \wedge \ldots \wedge \mathbf{U}p_n) \rightarrow BOOL(\mathbf{E}(p_i \wedge \mathbf{M}p_j), \mathbf{E}(p_i \wedge p_j))$$

where the operator $\mathbf{U}$ (at a *unique* world/moment) is defined as

$$\mathbf{U}\varphi := \mathbf{E}(\varphi \wedge \neg\mathbf{D}\varphi). \qquad \blacksquare$$

As a corollary it follows that for example

asymmetry: $\qquad\qquad\ \forall\,xy(x < y \rightarrow \neg\,y < x)$ and
almost-connectedness: $\forall\,xyz(x < y \rightarrow (x < z \vee z < y))$

are also definable. Also more complicated first-order properties such as discreteness (see above) become definable. However, we can still not define *all* first-order properties as is witnessed by the following proposition.

Proposition 4.2.4 The existence of a reflexive world ($\exists w\ wRw$) is not definable in $\text{PML}(R, \neq)$.

Proof: Although most previous preservation results are invalid now, we can use the filtration method (see Definition 3.2.10 in section 2 of Chapter 3) as follows. Suppose that φ defines the existence of a reflexive world, then

φ is refuted on $(\mathsf{IN}, <)$ (IN is the set of natural numbers). So we can find a valuation V such that

$$\mathcal{M} \not\models \varphi \quad \text{for} \quad \mathcal{M} = (\mathsf{IN}, <, V).$$

We are going to apply filtration to $\mathcal{M}$ and φ. So, let Ψ be the finite set consisting of φ together with all its subformulas and define for each $n \in \mathsf{IN}$:

$$\Psi(n) := \{\psi \in \Psi \mid \mathcal{M}, n \models \psi\}.$$

Since Ψ is finite the $\Psi(n)$ partition IN into a finite number of classes. Hence, a certain number of these classes, say k $(k > 0)$, occur infinitely often and there exists $N \in \mathsf{IN}$ such that from N onwards only these classes occur. Let us denote these classes that correspond to an infinite subset of IN by $\Psi_1, \ldots, \Psi_k$. Now, our filtrated model $\mathcal{M}_\Psi = (W_\Psi, R_\Psi, V_\Psi)$ is not standard (see the remark after this proof) but has some special properties. It consists of $N + 2 \cdot k$ worlds with the following connection between the old and the new worlds. The first N worlds correspond to $0, \ldots, N-1$ without any change. For $n \geq N$, n corresponds with Ψ_i, $1 \leq i \leq k$, such that $\Psi(n) = \Psi_i$ and with a duplicate Ψ_i' of Ψ_i. The $2 \cdot k$ worlds $\Psi_1, \Psi_1', \ldots, \Psi_k, \Psi_k'$ form a cluster, that is, they are all R_Ψ-related. By induction one easily establishes (as for standard filtration) that for all $n \in \mathsf{IN}$ and all $\psi \in \Psi$: ψ holds in $\mathcal{M}$ at n iff ψ holds at the corresponding world(s) in $\mathcal{M}_\Psi$. But then it follows that φ is refuted on $\mathcal{M}_\Psi$, a finite model with reflexive worlds, a contradiction. Hence, such a φ defining the existence of a reflexive world cannot exist.   ■

Remark 4.2.5 The role of the duplicates $\Psi_1', \ldots, \Psi_k'$ relates to the presence of the **D**-operator. Because of this operator the standard filtration technique does not work any more. Take for example the formula **D**⊤, then standard filtration will collapse every infinite model into a single world which obviously is not truth-preserving since **D**⊤ will not hold in this filtrated model. Therefore we need to double worlds which correspond to more than one world in the original model. The induction proof that corresponding (doubled) points verify the same formulas (in the above proof restricted to the set Ψ) will reappear in a more elaborate form in the completeness proofs of section 3.

Apart from conditions on the alternative (precedence) relation, the addition of operators for inequality also allows properties of the set of worlds (moments) to be defined as in the following proposition.

Proposition 4.2.6 Every finite cardinality is definable in PML($\neq$).

Proof: For all $n \in \mathbb{N}$, $|W| \leq n$ is a universal first-order condition on $=$ and hence is definable in $\mathrm{PML}(\neq)$ by the proof of Theorem 4.2.3. Furthermore, $|W| > n$ is defined by

$$\mathbf{A} \bigvee_{i=1}^{n} p_i \rightarrow \mathbf{E} \bigvee_{i=1}^{n} (p_i \wedge \mathbf{D}p_i).$$

We prove this equivalence as follows.

First suppose that $|W| > n$. Then we can choose $n + 1$ different worlds $w_1, \ldots, w_{n+1}$. Consider any valuation V verifying $\mathbf{A} \bigvee_{i=1}^{n} p_i$, then for each j, $1 \leq j \leq n + 1$, at least one of the p_i's ($1 \leq i \leq n$) is true at w_j. By the pigeonhole principle this implies that there exist j, $1 \leq j \leq n + 1$, and j', $1 \leq j' \leq n + 1$, with $j' \neq j$ such that p_i is true both at w_j and at $w_{j'}$ for some i, $1 \leq i \leq n$. So, $p_i \wedge \mathbf{D}p_i$ is true at w_j and hence $\mathbf{E} \bigvee_{i=1}^{n} (p_i \wedge \mathbf{D}p_i)$ holds at arbitrary worlds. Thus, $\mathbf{A} \bigvee_{i=1}^{n} p_i \rightarrow \mathbf{E} \bigvee_{i=1}^{n} (p_i \wedge \mathbf{D}p_i)$ holds at arbitrary worlds for all valuations V.

Conversely, suppose that $\mathbf{A} \bigvee_{i=1}^{n} p_i \rightarrow \mathbf{E} \bigvee_{i=1}^{n} (p_i \wedge \mathbf{D}p_i)$ holds for all valuations V but that $|W| \leq n$, say $W \subseteq \{w_1, \ldots, w_n\}$. Then, the particular valuation V assigning $\{w_i\}$ to p_i for those i, $1 \leq i \leq n$, such that $w_i \in W$ and $\emptyset$ to the other p_i's ($1 \leq i \leq n$), verifies $\mathbf{A} \bigvee_{i=1}^{n} p_i$. Consequently, V verifies also $\mathbf{E} \bigvee_{i=1}^{n} (p_i \wedge \mathbf{D}p_i)$, but this asserts the existence of j, $1 \leq j \leq n$, and j', $1 \leq j' \leq n$, and i, $1 \leq i \leq n$, such that $w_j \in W$ and $w_{j'} \in W$ and $j \neq j'$ and p_i holds both at w_j and at $w_{j'}$. Since $|V(p_i)| \leq 1$ for all i, $1 \leq i \leq n$, this is impossible. Hence it must be the case that $|W| > n$.

$|W| = n + 1$ can then be defined by a conjunction of $|W| \leq n + 1$ and $|W| > n$.

∎

Because of filtration, infinity of W can obviously not be defined. In fact, no essentially higher-order property of identity can be defined:

Proposition 4.2.7 All formulas from $\mathrm{PML}(\neq)$ define first-order conditions over identity $=$.

Proof: Formulas from $\mathrm{PML}(\neq)$ translate into the monadic second-order logic over pure identity and all formulas of this second-order logic are equivalent with first-order formulas (see [Ack 62]).

∎

On the other hand, all first-order formulas over identity can be defined as a Boolean combination of formulas expressing the existence of at least a certain number of elements. Since the latter formulas are definable in PML($\neq$) by the proof of Proposition 4.2.6 it follows that

Corollary 4.2.8 Over frames PML($\neq$) is equivalent with first-order logic over $=$.

Another interesting topic related to expressive power considerations concerns the possibility to discriminate between special structures. For example, ordinary modal logic cannot discriminate between $\mathbb{N}$ and $\mathbb{Z}$, the set of natural numbers and integers, respectively (see Proposition 3.2.23 in section 2 of Chapter 3). Adding inequality again helps:

Proposition 4.2.9 ([Rijk 89]) There exists $\varphi \in$ PML($R, \neq$) such that $(\mathbb{Z}, <) \models \varphi$, but $(\mathbb{N}, <) \not\models \varphi$.

Proof: This follows from the fact that the existence of a (different) predecessor is definable by $p \rightarrow \mathbf{DM}p$. ∎

For two finite frames it follows from the theorem about definability of universal first-order conditions (Theorem 4.2.3) that they are isomorphic if and only if they validate the same formulas from PML($R, \neq$):

Proposition 4.2.10 ([Rijk 89]) For finite frames $\mathcal{F}$ and $\mathcal{F}'$:

$$\mathcal{F} \cong \mathcal{F}' \quad \text{iff} \quad (\mathcal{F} \models \varphi \Leftrightarrow \mathcal{F}' \models \varphi \text{ for all } \varphi \in \text{PML}(R, \neq)).$$

Proof: Finite frames are isomorphic if and only if they have the same universal first-order theory. By Theorem 4.2.3 this universal first-order theory is definable in PML($R, \neq$). ∎

For temporal logic examples will necessarily be more difficult because the **D**-operator is expressible over linear orders: $\mathbf{D}\varphi \equiv \mathbf{P}\varphi \vee \mathbf{F}\varphi$. The topic of the characterization of special structures is an interesting one and deserves further investigation.

A related topic is that of categoricity, see Definition 3.2.24 in section 2 of Chapter 3. That section also contains two propositions (Propositions 3.2.25 and 3.2.26) showing that the notions of categoricity and of ω-categoricity are rather meaningless for modal and temporal logic. This is however not the case for **D**-logics: unlike modal and temporal logic, **D**-logics have categorical theories. For example, the theory of a finite frame $\mathcal{F}$ is categorical.

This follows from the propositions about definability of finite cardinalities (Proposition 4.2.6) and about isomorphic finite frames (Proposition 4.2.10): suppose $\mathcal{F}'$ validates the theory of $\mathcal{F}$, then $\mathcal{F}'$ must have the same number of elements as $\mathcal{F}$ has by Proposition 4.2.6, and by Proposition 4.2.10 then $\mathcal{F} \cong \mathcal{F}'$ because $\mathcal{F}$ and $\mathcal{F}'$ validate the same formulas. Another example of categoricity is provided by:

Proposition 4.2.11 ([Rijk 89]) The complete $\text{PML}(<,>,\neq)$-theory of $\mathbb{Z}$ is categorical.

Proof: It can be shown that $\mathbb{Z}$ is definable in $\text{PML}(<,>)$, that is, pure temporal logic, on the class of all strict linear orderings (in fact, on the class of connected strict partial orderings). Since this class is defined by universal first-order sentences it follows from Theorem 4.2.3 that $\mathbb{Z}$ is definable in $\text{PML}(<,>,\neq)$. ■

Also the notion of ω-categoricity makes sense for **D**-logics, for example:

Proposition 4.2.12 ([Rijk 89]) The complete $\text{PML}(R,\neq)$-theory of $\mathbb{Q}$ is ω-categorical.

Proof: By the previous results on the definability of first-order conditions, we can give formulas of $\text{PML}(R,\neq)$ equivalent (on frames) to the axioms for the theory of dense linear orderings without endpoints. As is well known, any countably infinite dense linear ordering without endpoints is isomorphic to $\mathbb{Q}$. The requested formulas are:

TRANS $\mathbf{L}\,\mathbf{L}\,p \;\rightarrow\; \mathbf{L}\,p$

IRREF $\mathbf{M}\,p \;\rightarrow\; \mathbf{D}\,p$

LIN $(\mathbf{U}\,p \wedge \mathbf{U}\,q) \;\rightarrow\; (\mathbf{E}(p \wedge q) \vee \mathbf{E}(p \wedge \mathbf{M}\,q) \vee \mathbf{E}(q \wedge \mathbf{M}\,p))$

DENS $\mathbf{L}\,p \;\rightarrow\; \mathbf{L}\,\mathbf{L}\,p$

SUC-P $\mathbf{U}\,p \;\rightarrow\; \mathbf{D}\,\mathbf{M}\,p$

SUC-F $\mathbf{M}\,\top$ ■

The other question about the correspondence of modal and first-order formulas asked which modal formulas are definable by a first-order property over *frames*. Section 2 of Chapter 3 contains a theorem (Theorem 3.2.28) that states the first-order definability of all Sahlqvist-forms. This syntactical class can easily be redefined for polymodal logics such as $\text{PML}(R,\neq)$ as follows.

Definition 4.2.13 A formula of PML($R, \neq$) is called a *Sahlqvist-form* when it is of the form $\varphi \to \psi$ where

  (i) φ is constructed from $p, \mathbf{L}p, \mathbf{LL}p, \ldots, \overline{\mathbf{D}}p, \overline{\mathbf{D}}\,\overline{\mathbf{D}}p, \ldots \perp, \top$ using only $\land, \lor, \mathbf{M}$ and $\mathbf{D}$, while

 (ii) ψ is constructed from proposition letters, $\perp, \top$ using $\land, \lor, \mathbf{M}, \mathbf{D}, \mathbf{L}$ and $\overline{\mathbf{D}}$.

Thus, instead of $\mathbf{M}$ we may also use $\mathbf{D}$, and similarly for $\mathbf{L}$ and $\overline{\mathbf{D}}$. The definition for other polymodal logics such as PML($<, >, \neq$) is similar. Again (see Remark 3.2.29 in section 2 of Chapter 3) this class of formulas is not as restrictive as it appears at first sight. For example, the translations of universal first-order conditions in the proof of Theorem 4.2.3 can be rewritten as Sahlqvist-forms. Also for the new definition of Sahlqvist-forms we have:

Theorem 4.2.14 All Sahlqvist-forms define first-order conditions.

Proof: The corresponding theorem (Theorem 3.2.28) in section 2 of Chapter 3 can easily be generalized to polymodal logics. Again we demonstrate by an example that all Sahlqvist-forms are locally equivalent with a first-order condition containing precisely one free variable x. For this purpose we use the formula of PML($<, >, \neq$) defining comparability:

$$\mathbf{D}\,p \;\to\; (\mathbf{P}\,p \lor \mathbf{F}\,p).$$

This formula is already a Sahlqvist-form so that we do not need to transform it into one. Application of the standard translation (with of course the adaptation for the $\mathbf{D}$-operator as given earlier in this section) gives the first-order formula

$$\exists\,y\,(x \neq y \land Py) \;\to\; (\exists\,z\,(z < x \land Pz) \lor \exists\,z'\,(x < z' \land Pz')).$$

Again we rewrite the existential quantification of the antecedent as a universal quantification over the whole formula:

$$\forall\,y\,((x \neq y \land Py) \;\to\; (\exists\,z\,(z < x \land Pz) \lor \exists\,z'\,(x < z' \land Pz'))).$$

Take u as a variable that does not occur in this formula. The antecedent $\mathbf{D}p$ of the Sahlqvist-form contains only one proposition letter, namely p. So we get for the method of substitutions:

$$v(p) = y$$
$$\text{and}$$
$$CV(p, \mathbf{D}p): \ y = u.$$

In the first-order formula above we then have to apply the following substitutions:

$$[y/u]CV(p, \mathbf{D}p) \quad \text{for } Py,$$
$$[z/u]CV(p, \mathbf{D}p) \quad \text{for } Pz,$$
$$[z'/u]CV(p, \mathbf{D}p) \quad \text{for } Pz'.$$

This yields the first-order formula

$$\forall\, y\, ((x \neq y \ \wedge\ y = y) \rightarrow (\exists\, z\, (z < x \ \wedge\ y = z)\ \vee\ \exists\, z'\, (x < z' \ \wedge\ y = z'))).$$

After simplification this becomes

$$\forall\, y\, (x \neq y \ \rightarrow\ (y < x \ \vee\ x < y))$$

whose global version (obtained by prefixing with $\forall x$) is indeed comparability:

$$\forall\, xy\, (x = y \ \vee\ y < x \ \vee\ x < y). \qquad\blacksquare$$

We conclude our semantic survey by an observation relating $\mathbf{E}$ and $\mathbf{A}$.

Proposition 4.2.15 Let $\mathcal{M} = (T, <, V)$ be a (temporal) model. Then $\mathcal{M} \models \mathbf{E}\varphi \rightarrow \mathbf{A}\varphi$ for all φ if and only if

(i) $<$ satisfies either

    (a) both succession towards past and succession towards future, or

    (b) *all* points are both a beginning and an end (i.e. there are only isolated points),

(ii) V is uniform, that is, for all proposition letters p either $V(p) = \emptyset$ or $V(p) = T$.

Proof: The case $|T| \leq 1$ is trivial, so suppose $|T| > 1$. First we treat the only if case, so suppose $\mathcal{M} \models \mathbf{E}\varphi \rightarrow \mathbf{A}\varphi$ for all φ. Then (ii) follows immediately by taking $\varphi \equiv p$. (i) follows by observing that either

(a) $\mathcal{M} \models \mathbf{EP}\top$ and hence also $\mathcal{M} \models \mathbf{EF}\top$, so by taking $\varphi \equiv \mathbf{P}\top$ and $\varphi \equiv \mathbf{F}\top$ we get $\mathcal{M} \models \mathbf{AP}\top$ (SUC-P) and $\mathcal{M} \models \mathbf{AF}\top$ (SUC-F) *or*

(b) $\mathcal{M} \not\models \mathbf{EP}\top$ and hence also $\mathcal{M} \not\models \mathbf{EF}\top$, so $\mathcal{M} \models \mathbf{AH}\bot$ (all points are a beginning) and $\mathcal{M} \models \mathbf{AG}\bot$ (all points are an end).

For the if case, suppose $\mathcal{M}$ satisfies (i) and (ii). That $\mathcal{M} \models \mathbf{E}\varphi \to \mathbf{A}\varphi$ for all φ is proved by induction. (ii) gives the basic induction step for proposition letters. Case (ib) is easy (this case can also be proved by a symmetry argument). So suppose we have to deal with (ia). A typical case is $\varphi \equiv \mathbf{F}\psi$:

$$\mathcal{M} \models \mathbf{EF}\psi \Rightarrow \mathcal{M} \models \mathbf{E}\psi \Rightarrow \mathcal{M} \models \mathbf{A}\psi \Rightarrow \mathcal{M} \models \mathbf{AF}\psi$$

where the one but last step is justified by the induction hypothesis and the last step by SUC-F ($\mathcal{M} \models \mathbf{AF}\top$). The case $\varphi \equiv \mathbf{D}\psi$ is similar but now the last step $\mathcal{M} \models \mathbf{A}\psi \Rightarrow \mathcal{M} \models \mathbf{AD}\psi$ is justified by $|T| \neq 1$ and hence $\mathcal{M} \models \mathbf{AD}\top$.   ∎

4.3   Proof Theory

We now turn to axiomatizations of polymodal logics with inequality. First we present complete proof systems for the basic logics $\mathrm{PML}(\neq)$, $\mathrm{PML}(R, \neq)$ and $\mathrm{PML}(<, >, \neq)$.

Definition 4.3.1 The proof system D consists of a complete axiomatization of propositional logic including the rule of Modus Ponens (see section 2 of Chapter 3) and

(D2) $\overline{\mathbf{D}}\varphi := \neg\mathbf{D}\neg\varphi$

(R2) to infer $\overline{\mathbf{D}}\varphi$ from φ

(A1) $\overline{\mathbf{D}}(\varphi \to \psi) \to (\overline{\mathbf{D}}\varphi \to \overline{\mathbf{D}}\psi)$

(A2) $\varphi \to \overline{\mathbf{D}}\mathbf{D}\varphi$ (symmetry)

(A3) $\mathbf{DD}\varphi \to (\varphi \vee \mathbf{D}\varphi)$ (pseudo-transitivity).

The completeness proof of D uses the following theorem of D.

Proposition 4.3.2 $\vdash_D \mathbf{D}(\varphi \wedge \psi) \to \mathbf{D}\varphi$.

Proof: This theorem of D can be derived as follows.

1. $\neg\varphi \to \neg(\varphi \wedge \psi)$ (propositional logic)

2. $\overline{\mathbf{D}}(\neg\varphi \to \neg(\varphi \wedge \psi))$ (1,R2)

3. $\overline{\mathbf{D}}\neg \varphi \;\to\; \overline{\mathbf{D}}\neg(\varphi \wedge \psi)$ (2,A1,Modus Ponens)

4. $\neg \overline{\mathbf{D}} \neg(\varphi \wedge \psi) \;\to\; \neg \overline{\mathbf{D}} \neg \varphi$ (3,propositional logic)

5. $\mathbf{D}(\varphi \wedge \psi) \;\to\; \mathbf{D}\varphi$ (4,D2)

$\blacksquare$

Theorem 4.3.3 (Completeness of D)   For all $\varphi \in \mathrm{PML}(\neq)$ and $\Psi \subseteq$ $\mathrm{PML}(\neq)$:

$$\Psi \vdash_D \varphi \quad \text{if and only if} \quad \Psi \models_m \varphi.$$

Proof: Soundness is standard by induction on the length of derivations. As an example we show that the new rule (R2) preserves validity and that the new axiom schemas (A1)–(A3) are valid. To start with rule (R2), suppose that φ is valid, then for all models $\mathcal{M}$ and all worlds w: $\mathcal{M}, w \models \varphi$. We have to show that $\mathcal{M}, w \models \overline{\mathbf{D}}\varphi$ for all models $\mathcal{M}$ and all worlds w. So, taking $w' \neq w$, we have to show $\mathcal{M}, w' \models \varphi$ which follows immediately from the hypothesis that φ is valid.

To check axiom schema (A1), we have to show that for all formulas φ and ψ, all models $\mathcal{M}$ and worlds w: $\mathcal{M}, w \models \overline{\mathbf{D}}(\varphi \to \psi) \to (\overline{\mathbf{D}}\varphi \to \overline{\mathbf{D}}\psi)$. This reduces to: suppose $\mathcal{M}, w \models \overline{\mathbf{D}}(\varphi \to \psi)$ and $\mathcal{M}, w \models \overline{\mathbf{D}}\varphi$, prove that $\mathcal{M}, w \models \overline{\mathbf{D}}\psi$. Well, $\mathcal{M}, w \models \overline{\mathbf{D}}(\varphi \to \psi)$ means that for all $w' \neq w$: $\mathcal{M}, w' \models \varphi$ implies $\mathcal{M}, w' \models \psi$. The second hypothesis $\mathcal{M}, w \models \overline{\mathbf{D}}\varphi$ means that for all $w' \neq w$ $\mathcal{M}, w' \models \varphi$. The conclusion that for all $w' \neq w$ $\mathcal{M}, w' \models \psi$ is immediate.

To check that axiom schema (A2) is valid, we have to show that for all formulas φ, all models $\mathcal{M}$ and all worlds w: $\mathcal{M}, w \models \varphi \to \overline{\mathbf{D}}\mathbf{D}\varphi$. This reduces to: if $\mathcal{M}, w \models \varphi$ then $\mathcal{M}, w \models \overline{\mathbf{D}}\mathbf{D}\varphi$. So, supposing $\mathcal{M}, w \models \varphi$, take $w' \neq w$. To prove that $\mathcal{M}, w' \models \mathbf{D}\varphi$, that is, that there exists $w'' \neq w'$ so that $\mathcal{M}, w'' \models \varphi$. By taking $w'' = w$ this follows immediately from the hypothesis.

To check axiom schema (A3), we have to show that for all formulas φ, all models $\mathcal{M}$ and all worlds w: $\mathcal{M}, w \models \mathbf{D}\mathbf{D}\varphi \to (\varphi \vee \mathbf{D}\varphi)$. So, suppose that $\mathcal{M}, w \models \mathbf{D}\mathbf{D}\varphi$. Then there exists $w' \neq w$ such that $\mathcal{M}, w' \models \mathbf{D}\varphi$ and furthermore $w'' \neq w'$ such that $\mathcal{M}, w'' \models \varphi$. Now either $w'' = w$ in which case $\mathcal{M}, w \models \varphi$, or $w'' \neq w$ in which case $\mathcal{M}, w \models \mathbf{D}\varphi$. These two possibilities lead to the desired conclusion $\mathcal{M}, w \models \varphi \vee \mathbf{D}\varphi$.

For the proof of adequacy we use the same techniques as in the completeness proof of the minimal modal logic proof system K (see Theorem 3.2.31 in section 2 of Chapter 3). So suppose $\Psi \not\vdash_D \psi_0$. To prove that $\Psi \not\models_m \psi_0$. Let $\mathcal{M}_0$ be the standard Henkin Model of all maximally Ψ-consistent sets of formulas from PML($\neq$) with a relation $\not\approx$ defined by

$$\Phi_1 \not\approx \Phi_2 \quad \text{if, for all formulas } \varphi, \; \overline{D}\varphi \in \Phi_1 \; \Rightarrow \; \varphi \in \Phi_2.$$

In the sequel we also use the equivalent formulation

$$\Phi_1 \not\approx \Phi_2 \quad \text{if, for all formulas } \varphi, \; \varphi \in \Phi_2 \; \Rightarrow \; D\varphi \in \Phi_1.$$

This equivalence is easily shown as follows:

$$\text{for all formulas } \varphi, \; \overline{D}\varphi \in \Phi_1 \; \Rightarrow \; \varphi \in \Phi_2$$
$$\text{iff}$$
$$\text{for all formulas } \varphi, \; \varphi \notin \Phi_2 \; \Rightarrow \; \overline{D}\varphi \notin \Phi_1$$
$$\text{iff}$$
$$\text{for all formulas } \varphi, \; \neg\varphi \in \Phi_2 \; \Rightarrow \; D\neg\varphi \in \Phi_1$$
$$\text{iff}$$
$$\text{for all formulas } \varphi, \; \varphi \in \Phi_2 \; \Rightarrow \; D\varphi \in \Phi_1.$$

In the same way as in the completeness proof for K we can prove the Truth Lemma for $\mathcal{M}_0$ (where the new rule R2 and axiom schema A1 replace the Necessitation rule, respectively the Distribution axiom schema, both needed for the proof of the Truth Lemma). Now for arbitrary ψ such that $\Psi \not\vdash_D \psi$ (so in particular for ψ_0), $\{\neg\psi\}$ is Ψ-consistent, so by Lindenbaum's Lemma we can find a maximally Ψ-consistent set Φ_0 containing $\neg\psi$ and $\mathcal{M}_0, \Phi_0 \not\models \psi$ by the Truth Lemma. In the case of the proof system K, the proof of adequacy was complete at this point, because the semantics of PML(R), that is, ordinary modal logic, involves a binary relation R that is *arbitrary*. Consequently, the particular relation R_Ψ as defined for the Henkin Model in that completeness proof posed no problems. In the case of PML($\neq$) however, we are confronted with the very special relation of inequality in the semantics. In the soundness proof we showed that the new rule and axiom schemas were at least valid when interpreted over inequality. Adequacy, however, demands that whatever is valid for all models with inequality, can be derived in the proof system D. In the rest of the proof we mean by a standard model a model incorporating real inequality $\neq$. So far, we only constructed the model $\mathcal{M}_0$ with relation $\not\approx$ such that ψ_0 is refuted in $\mathcal{M}_0$. Our task is to construct a standard model out of $\mathcal{M}_0$ in which ψ_0 is refuted. To this end, let us first investigate which properties can already be ascribed to the relation $\not\approx$ of $\mathcal{M}_0$ because of the extra axiom schemas (A2) and (A3). These two schemas ensure that:

(i) $\forall\Phi_1\forall\Phi_2\ (\Phi_1\not\approx\Phi_2\ \Rightarrow\ \Phi_2\not\approx\Phi_1)$, respectively

(ii) $\forall\Phi_1\forall\Phi_2\forall\Phi_3\ ((\Phi_1\not\approx\Phi_2\ \text{and}\ \Phi_2\not\approx\Phi_3)\ \Rightarrow\ (\Phi_1=\Phi_3\ \text{or}\ \Phi_1\not\approx\Phi_3))$.

(so $\not\approx$ is symmetric and "pseudo-transitive").
We prove (i) and (ii) as follows:

(i) Suppose $\Phi_1\not\approx\Phi_2$. We have to show $\Phi_2\not\approx\Phi_1$ or that for all formulas φ: $\varphi\in\Phi_1\Rightarrow\mathbf{D}\varphi\in\Phi_2$. So let $\varphi\in\Phi_1$. By axiom schema (A2) it follows (since Φ_1 is maximally Ψ-consistent) that $\overline{\mathbf{D}}\mathbf{D}\varphi\in\Phi_1$. By the definition of $\Phi_1\not\approx\Phi_2$ the desired conclusion $\mathbf{D}\varphi\in\Phi_2$ is immediate.

(ii) Suppose $\Phi_1\not\approx\Phi_2,\Phi_2\not\approx\Phi_3$ and $\Phi_1\neq\Phi_3$. We have to show $\Phi_1\not\approx\Phi_3$ or that for all formulas φ: $\varphi\in\Phi_3\Rightarrow\mathbf{D}\varphi\in\Phi_1$. So let $\varphi\in\Phi_3$. Because $\Phi_1\neq\Phi_3$ and Φ_1,Φ_3 are maximally Ψ-consistent there exists a formula χ such that $\chi\in\Phi_3$ but $\chi\notin\Phi_1$. Now, since $\varphi\in\Phi_3,\chi\in\Phi_3$ and Φ_3 is maximally Ψ-consistent we have also $\varphi\wedge\chi\in\Phi_3$. So by $\Phi_2\not\approx\Phi_3$ it follows that $\mathbf{D}(\varphi\wedge\chi)\in\Phi_2$ and by $\Phi_1\not\approx\Phi_2$ furthermore that $\mathbf{D}\mathbf{D}(\varphi\wedge\chi)\in\Phi_1$. By axiom schema (A3) it follows (since Φ_1 is maximally Ψ-consistent) that $\varphi\wedge\chi\in\Phi_1$ or $\mathbf{D}(\varphi\wedge\chi)\in\Phi_1$. The first case is impossible because $\chi\notin\Phi_1$ and Φ_1 is maximally Ψ-consistent. Thus, $\mathbf{D}(\varphi\wedge\chi)\in\Phi_1$ and because $\mathbf{D}(\varphi\wedge\chi)\to\mathbf{D}\varphi$ is a theorem of D (see Proposition 4.3.2 preceding this completeness theorem) and Φ_1 is maximally Ψ-consistent the desired conclusion $\mathbf{D}\varphi\in\Phi_1$ is reached.

Our first improvement on model $\mathcal{M}_0$ to get a standard model is the smallest submodel of $\mathcal{M}_0$ containing Φ_0 and being closed under $\not\approx$, denoted by $\mathcal{M}_1$. Since $\mathcal{M}_1$ is a generated submodel of $\mathcal{M}_0$, it follows that (cf. the Generation Theorem in section 2 of Chapter 3, Theorem 3.2.6) for all formulas φ and all worlds (i.e. maximally Ψ-consistent sets) Φ from $\mathcal{M}_1$:

$$\mathcal{M}_1,\Phi\models\varphi\quad\text{if and only if}\quad\mathcal{M}_0,\Phi\models\varphi.$$

Our next claim is that $\not\approx$ holds between any two different points in $\mathcal{M}_1$:

$$\text{if}\ \ \Phi_0\not\approx^n\Phi_1\ \ \text{and}\ \ \Phi_0\not\approx^m\Phi_2,\ \text{then}\ \ \Phi_1\not\approx\Phi_2\ \ \text{or}\ \ \Phi_1=\Phi_2.$$

This follows from (i) and (ii) above: repetitive application of (ii) yields that $\Phi_0 \not\approx^n \Phi_1 \Rightarrow (\Phi_0 = \Phi_1$ or $\Phi_0 \not\approx \Phi_1)$ and similarly $\Phi_0 \not\approx^m \Phi_2 \Rightarrow (\Phi_0 = \Phi_2$ or $\Phi_0 \not\approx \Phi_2)$.
Differentiate between three cases:

(1) $\Phi_0 = \Phi_1$: substituting this in the second implication above immediately gives the desired conclusion $\Phi_1 = \Phi_2$ or $\Phi_1 \not\approx \Phi_2$

(2) $\Phi_0 = \Phi_2$: substituting this in the first implication and applying (i) yields again the desired conclusion $\Phi_1 = \Phi_2$ or $\Phi_1 \not\approx \Phi_2$.

(3) $\Phi_0 \not\approx \Phi_1$ and $\Phi_0 \not\approx \Phi_2$: by (i) $\Phi_1 \not\approx \Phi_0$ and together with $\Phi_0 \not\approx \Phi_2$ it follows from (ii) that $\Phi_1 = \Phi_2$ or $\Phi_1 \not\approx \Phi_2$.

So, at least we achieved in $\mathcal{M}_1$ that

$$\Phi_1 \neq \Phi_2 \;\Rightarrow\; \Phi_1 \not\approx \Phi_2$$

for all worlds Φ_1 and Φ_2.

Therefore, the only 'non-standard' feature of $\not\approx$ left when compared to real inequality is the possibility of reflexive worlds Φ, that is, where $\Phi \not\approx \Phi$ holds. Now, let $\mathcal{M}_2$ be the model that replaces each $\not\approx$-reflexive point Φ of $\mathcal{M}_1$ by two points Φ', Φ'' such that $\Phi' \not\approx \Phi''$ and $\Phi'' \not\approx \Phi'$ and all $\not\approx$-connections to other points are maintained and Φ', Φ'' have the same valuation as Φ. Our last claim is that for all formulas φ and all worlds Φ of $\mathcal{M}_2$:

$$\mathcal{M}_2, \Phi \models \varphi \quad \text{if and only if} \quad \mathcal{M}_1, \tilde{\Phi} \models \varphi$$

where $\tilde{\Phi} = \Phi$ if Φ was not $\not\approx$-reflexive and $\widetilde{\Phi'} = \widetilde{\Phi''} = \Phi$ for (doubled) $\not\approx$-reflexive points Φ. This claim is proved by induction on φ:

(a) $\varphi \equiv p$ is immediate since Φ and $\tilde{\Phi}$ have the same valuation

(b) the cases $\varphi \equiv \neg\varphi_1$ and $\varphi \equiv \varphi_1 \wedge \varphi_2$ are immediate from the induction hypothesis

(c) $\varphi \equiv \overline{D}\psi$: To prove: $\mathcal{M}_2, \Phi \models \overline{D}\psi$ if and only if $\mathcal{M}_1, \tilde{\Phi} \models \overline{D}\psi$.

   (c1) only if: easy since each world Φ_1 of $\mathcal{M}_1$ can be written as $\widetilde{\Phi_2}$ for a world Φ_2 of $\mathcal{M}_2$

(c2) if: in case Φ was not $\not\approx$-reflexive this follows immediately from the induction hypothesis; otherwise $\tilde{\Phi} \not\approx \tilde{\Phi}$, hence $\mathcal{M}_1, \tilde{\Phi} \models \overline{\mathbf{D}}\psi$ implies $\mathcal{M}_1, \tilde{\Phi} \models \psi$, so by the induction hypothesis $\mathcal{M}_2, \Phi' \models \psi$ and $\mathcal{M}_2, \Phi'' \models \psi$.

$\mathcal{M}_2$ is a standard model (with real inequality $\neq$) where ψ_0 is refuted, as required.

∎

Definition 4.3.4 The proof system D_m consists of the minimal modal logic proof system K (see Definition 3.2.30 in section 2 of Chapter 3) together with the above system D (see Definition 4.3.1) plus the axiom schema

$$\mathbf{M}\varphi \;\rightarrow\; (\varphi \vee \mathbf{D}\varphi) \hspace{4cm} \text{(relation M and D)}.$$

Theorem 4.3.5 (Completeness of D_m) For all $\varphi \in \text{PML}(R, \neq)$ and $\Psi \subseteq \text{PML}(R, \neq)$:

$$\Psi \vdash_{D_m} \varphi \quad \text{if and only if} \quad \Psi \models_m \varphi.$$

Proof: The proof above can easily be adapted. The additional axiom schema ensures that

$$\forall \Phi \; \forall \Phi' \; (\Phi \; R \; \Phi' \;\Rightarrow\; (\Phi' = \Phi \;\text{ or }\; \Phi \not\approx \Phi')).$$

Therefore closure under R remains within the closure under $\not\approx$, so we can use the previous construction.

∎

Definition 4.3.6 The proof system D_t consists of the minimal temporal logic proof system K_t (see Definition 3.2.32 in section 2 of Chapter 3) together with the proof system D plus the two axiom schemas

$$\mathbf{F}\varphi \;\rightarrow\; (\varphi \vee \mathbf{D}\varphi)$$
$$\mathbf{P}\varphi \;\rightarrow\; (\varphi \vee \mathbf{D}\varphi).$$

Theorem 4.3.7 (Completeness of D_t) For all $\varphi \in \text{PML}(<, >, \neq)$ and $\Psi \subseteq \text{PML}(<, >, \neq)$:

$$\Psi \vdash_{D_t} \varphi \quad \text{if and only if} \quad \Phi \models_m \varphi.$$

Proof: As in the previous proof. The additional axiom schemas now guarantee

$$\forall \Phi \, \forall \Phi' \, (\Phi < \Phi' \, \Rightarrow \, (\Phi' = \Phi \, \text{ or } \, \Phi \not\approx \Phi')) \quad \text{and}$$
$$\forall \Phi \, \forall \Phi' \, (\Phi' < \Phi \, \Rightarrow \, (\Phi' = \Phi \, \text{ or } \, \Phi \not\approx \Phi')). \qquad \blacksquare$$

Remark 4.3.8 Notice that we did not impose special restrictions on temporal frames, in particular we do not assume that $<$ is irreflexive. In the case that we restrict ourselves to irreflexive frames the above axiom schemas should be strengthened into $\mathbf{F}\varphi \to \mathbf{D}\varphi$ and $\mathbf{P}\varphi \to \mathbf{D}\varphi$.

Using filtration it follows that these logics satisfy the finite model property and hence are decidable (see section 2 of Chapter 3).

After having looked at completeness results for $\mathrm{PML}(\neq)$, $\mathrm{PML}(R, \neq)$ and $\mathrm{PML}(<, >, \neq)$ with respect to general classes of models we now give complete proof systems for three special classes of models, viz. linear orderings, dense linear orderings without endpoints and models isomorphic to the integers. These results are from [Rijk 89].

Definition 4.3.9 The proof system D_{lin} consists of the proof system D_m plus the three axiom shemes:

$\mathbf{M}\varphi \; \to \; \mathbf{D}\varphi$	(irreflexivity)
$\mathbf{MM}\varphi \; \to \; \mathbf{M}\varphi$	(transitivity)
$\varphi \; \to \; \mathbf{M}\psi \vee \overline{\mathbf{D}}(\psi \; \to \; \mathbf{M}\varphi)$	(comparability).

D_{lin} is complete with respect to linear orderings:

Theorem 4.3.10 For all $\varphi \in \mathrm{PML}(R, \neq)$ and $\Psi \subseteq \mathrm{PML}(R, \neq)$:

$$\Psi \vdash_{D_{lin}} \varphi \quad \text{if and only if} \quad \text{for any linearly ordered model } \mathcal{M},$$
$$\text{if } \mathcal{M} \models \Psi, \text{then } \mathcal{M} \models \varphi.$$

Proof: We adapt the proofs of completeness of D_m and D as follows. Instead of doubling $\not\approx$-reflexive points at the end of the completeness proof of D, we now replace each such point by a copy of IN with its standard ordering and with real inequality and the same valuation as the original point and the proper relations with the other points. It can again be proved by structural induction that the new and old model validate the same formulas. $\blacksquare$

Definition 4.3.11 The proof system D_{rat} consists of the proof system D_{lin} plus the three axiom schemes:

MⳆ   (successiveness to the right)

$\varphi \rightarrow DM\varphi$ (successiveness to the left)

$M\varphi \rightarrow MM\varphi$ (denseness).

D_{rat} is complete with respect to dense linear orderings without endpoints:

Theorem 4.3.12 For all $\varphi \in PML(R, \neq)$ and $\Psi \subseteq PML(R, \neq)$:

$\Psi \vdash_{D_{rat}} \varphi$ if and only if

for any dense linearly ordered model $\mathcal{M}$ without endpoints,

if $\mathcal{M} \models \Psi$, then $\mathcal{M} \models \varphi$.

Proof: As in the previous proof, but now replacing $\neq$-reflexive points by a copy of $\mathbb{Q}$, etcetera.   ■

Definition 4.3.13 The proof system D_{int} is obtained from the proof system D_{rat} by deleting the axiom scheme for denseness and adding the following axiom scheme:

$$L(L\varphi \rightarrow \varphi) \rightarrow (ML\varphi \rightarrow L\varphi).$$

Theorem 4.3.14 For all $\varphi \in PML(R, \neq)$ and *finite* $\Psi \subseteq PML(R, \neq)$:

$\Psi \vdash_{D_{int}} \varphi$ if and only if   for all models $\mathcal{M}$ isomorphic to $(\mathbb{Z}, <)$,

if $\mathcal{M} \models \Psi$, then $\mathcal{M} \models \varphi$.

The restriction to finite Ψ in the theorem above is necessary because compactness fails on models isomorphic to $(\mathbb{Z}, <)$. The proof of this theorem resorts to a different method of proving completeness and can be found as the proof of Theorem 4.12 in [Rijk 89]. More completeness results on special classes of models such as (strict) partial orders are contained in section 3 of [Rijk 90].

After having presented these complete axiomatizations of $\models_m$ we now look for similar results for $\models_f$. As for modal and temporal logic (see section 2 of Chapter 3) we can only obtain such results for special Φ. So we search for Φ that are frame-complete, that is, Φ such that for all φ

$$\Phi \vdash^s \varphi \text{ iff } \forall \mathcal{F}(\mathcal{F} \models \Phi \Rightarrow \mathcal{F} \models \varphi),$$

where $\vdash^s$ stands for one of the above proof systems with an additional rule of substitution that allows to infer any substitution instance of a formula already obtained. Φ containing only valuation-independent (closed) formulas (i.e. formulas without any proposition letters) such as the formulas defining BEGIN, END, SUC-P and SUC-F in section 2 can easily be proved frame-complete as follows. For a closed formula φ we have for all frames $\mathcal{F}$ and all valuations V:

$$\mathcal{F} \models \varphi \Leftrightarrow (\mathcal{F}, V) \models \varphi.$$

From this it is easy to prove for Φ only containing closed formulas that for all ψ

$$\Phi \models_f \psi \Leftrightarrow \Phi \models_m \psi.$$

By the above completeness theorems for $\models_m$ it then follows that Φ is frame-complete. In this way combinations of BEGIN, END, SUC-P and SUC-F yield 8 completeness theorems (the pairs BEGIN, SUC-P and END, SUC-F are mutually exclusive).

We can also obtain more general completeness results for frames, for example:

Proposition 4.3.15 When φ corresponds to a frame-condition α purely on $<$ and α also holds in the underlying frame of the standard Henkin Model, then $\{\varphi\}$ is frame-complete (this includes all φ that correspond to universal conditions α).

Proof: The reason is this: when inspecting the completeness proofs above we observe that the doubling of $\not\approx$-reflexive points to get a model with real inequality gives a surjective function F from the new model to the old one that is a *strong homomorphism*: $x < y$ iff $F(x) < F(y)$. The existence of such a function makes the new and old model elementary equivalent in the pure $<$-language. For these concepts from model theory the reader may consult [CK 73]. ∎

For conditions involving $<$ and $=$ it is not so easy to get such completeness results. For example, doubling $\not\approx$-reflexive points can disturb comparability $\forall xy(x < y \lor x = y \lor y < x)$. Nevertheless, we have a result for this case also.

Proposition 4.3.16 $\{\mathbf{D}\varphi \rightarrow (\mathbf{P}\varphi \lor \mathbf{F}\varphi) \mid \varphi \in \text{PML}(<, >, \neq)\}$ is frame-complete.

Proof: The given set is an axiom schema that enforces comparability on frames. Doubling $\not\approx$-reflexive points would disturb comparability. For a $\not\approx$-reflexive point x we use the following construction instead differentiating between two cases:

1. x is $<$-irreflexive. In this case just remove the $\not\approx$-loop in x: there is no change in evaluation because of the extra axiom schema $\mathbf{D}\varphi \rightarrow (\mathbf{P}\varphi \vee \mathbf{F}\varphi)$.

2. x is $<$-reflexive. In this case replace x by $(\mathbb{Z}, <)$, that is, the integers with their standard ordering, replacing $\not\approx$ by real inequality $\neq$ and using the same valuation everywhere.   ∎

Conjecture 4.3.17 The construction in the above proof is generalizable to a result stating completeness for all Sahlqvist-forms with respect to their corresponding first-order conditions.

On one hand this particular conjecture has been refuted in the meantime in section 5 of chapter 4 of [Rijk 89], on the other a similar conjecture for $PML(R, \neq)$ has been proven in section 3.3 of [Rijk 90].

A general question about completeness with respect to a class of frames (see Definition 3.2.34 in section 2 of Chapter 3) is the following: suppose that the pure temporal logic (i.e. based on the operators $\mathbf{F}$ and $\mathbf{P}$) of a class of frames is recursively axiomatizable, does the same hold for the temporal logic where inequality is added (i.e. based on the operators $\mathbf{F}$, $\mathbf{P}$ and $\mathbf{D}$)? And if so, can this be done via a uniform extension? For modal logic and $PML(R, \neq)$ such questions and partial answers are considered in section 3.4 of [Rijk 90].

4.4 Conclusions

We end this chapter with some conclusions. We extended modal and temporal logic with operators for reasoning about inequality. This simple idea has interesting consequences: all the usual first-order properties of the alternative and precedence relation are now definable. Furthermore, completeness and decidability results were given and several semantic results from modal and temporal logic could be adapted for the new logics. It is surprising that this simple idea has not been proposed before. However, ideas in a similar direction have been investigated independently in [Gor 88],[GG 89] and [Bla 89],[Bla 90]. In our terminology, [Gor 88] is concerned with the base language PML$(R, -R)$ where $-R$ denotes the complement of R. As an extension also the case PML$(R, -R, \neq, =)$ is briefly considered. Equality is easily axiomatized by $\mathbf{L}^= p \leftrightarrow p$ and the axiomatization of inequality is then derived from that given for complementary relations.

Like our idea to add an extra operator to modal and temporal logic to make these more expressive, the logic introduced in [Bla 89] is also motivated (although stemming from quite a different application area, viz. information systems) by expressive power considerations. However, the extension proposed in that paper uses additional variables, called nominals, instead of an additional operator for inequality. The resulting logic is called nominal tense logic. Its language consists of the Priorean propositional temporal logic of section 2 of Chapter 3 (with temporal operators $\mathbf{G}, \mathbf{F}, \mathbf{H}$ and $\mathbf{P}$) extended with nominals, represented by $i, i_1, \ldots, j, j_1, \ldots$, which are considered as atoms. The crucial point about nominals is that they are intended as propositions that are true at one and only one point. Therefore, the extension of the notion of a valuation (see Definition 3.2.1 in section 2 of Chapter 3) stipulates that for all nominals i, $V(i)$ is a singleton (instead of an arbitrary subset of the set of moments in case of normal propositions). So, nominals are so called because they name: they refer uniquely to points of time. Apart from this extension, the other semantic notions can be defined in the usual way. One of the main results of [Bla 89] is a complete axiomatization of nominal tense logic.

How is the expressive power of temporal logic affected by this addition of nominals? Like we did for PML$(<, >, \neq)$ in section 2 we give formulas defining first-order conditions that were not definable before:

IRREF: $i \rightarrow \neg \mathbf{F} i$ \hfill (irreflexivity)

LIN: $i \vee \mathbf{P} i \vee \mathbf{F} i$ \hfill (comparability).

Another indication of the obtained expressive power is given by the preservation results. For nominal tense logic it can be shown that preservation under disjoint unions and preservation under zigzag morphisms is lost, but that preservation under generated subframes and anti-preservation under ultrafilter extensions is maintained. The preservation result for generated subframes means for example that the existence of an isolated point (i.e. a point that is both a beginning and an end) cannot be defined (a counterexample is a frame with more than one point but exactly one isolated point: leaving out the isolated point gives a generated subframe). This is a difference with

$PML(<,>,\neq)$: as is clear from the defining formulas for a beginning and an end (see section 2), the existence of an isolated point can be defined by

$$\mathbf{E}\,(\,\mathbf{H}\perp\,\wedge\,\mathbf{G}\perp\,).$$

Indeed, for $PML(<,>,\neq)$ also preservation under generated subframes is lost and only anti-preservation under ultrafilter extensions remains. This gives rise to the question whether $PML(<,>,\neq)$ is strictly more expressive than nominal tense logic. The answer is positive and can be proved using the same techniques as for proving that all universal first-order conditions are definable (Theorem 4.2.3 in section 2):

Let $\varphi(i_1,\dots,i_n)$ be a formula of nominal tense logic whose nominals are $i_1,\dots,i_n$, then

$$\mathcal{F}\models\varphi(i_1,\dots,i_n)$$
$$\text{if and only if}$$
$$\mathcal{F}\models\mathbf{U}p_{i_1}\wedge\dots\wedge\mathbf{U}p_{i_n}\ \rightarrow\ [p_{i_1}/i_1,\dots,p_{i_n}/i_n]\,\varphi,$$

where $p_{i_1},\dots,p_{i_n}$ are propositions not occurring in φ.

In other words: propositions that are true at one and only one point (the intended function of nominals) can already be expressed in $PML(<,>,\neq)$ by the use of the uniqueness operator **U**.

The translation from nominal tense logic into $PML(<,>,\neq)$ also gives alternative ways of defining first-order conditions, for instance irreflexivity: the translation of $i\rightarrow\neg\mathbf{F}i$ is $\mathbf{U}p\rightarrow(p\rightarrow\neg\mathbf{F}p)$.

In [Bla 90] and [GG 89] the extension of the language $PML(R,W\times W)$, obtained by adding the **A**-operator to modal logic, with nominals is studied. A nice result in [GG 89] is that this language and $PML(R,\neq)$ are equally expressive, or in other words: adding inequality to modal logic gives the same additional expressive power as adding the universal relation and unique names. Also it turns out that the expressive power of $PML(R,W\times W)$ and modal logic with nominals is incomparable.

Another idea for using the **D**-operator is to add it to temporal logic with **until** and **since** operators (see section 3 of Chapter 3). Consider for example the closed (valuation-independent) formula

$$\mathbf{A}\,(\,\perp\ \textbf{until}\ \top\,).$$

This formula expresses a combination of discreteness and succession towards future.

Chapter 5

Message Passing Systems

5.1 Introduction

In this chapter we look at message passing systems and ways to specify them. First we describe the requirements which systems must fulfill in order to be qualified as a message passing system. Next we look at requirements for specification languages that are important in the context of message passing systems.

We will use temporal logic as a formalism for specifying message passing systems. Therefore, we first investigate the suitability of the standard temporal logics like those treated in Chapter 3 for this purpose. To that end we examine (propositional and first-order) temporal logics with **until** and **since** (as studied by Kamp and Stavi, see section 3 of Chapter 3) and their capability to specify certain classes of message passing systems. We prove that even such strong temporal logics (Kamp's logic is expressively complete with respect to the class of complete linear orders, and Stavi's extension makes it expressively complete with respect to the class of all linear orders) cannot express a large number of natural classes of message passing systems. This extends a result of Sistla et al. ([SCFG 82],[SCFM 84]) that unbounded buffers cannot be expressed in linear time temporal logic (a smaller class of message passing systems and a weaker logic). In our analysis the source of this inexpressiveness is the impossibility to *couple each message that is delivered by a message passing system to a* **unique** *message accepted by that system*. This result seems to necessitate the enrichment of TL-based formalisms for the specification of message passing systems, for instance with auxiliary data structures or histories as done, respectively, by Lamport and Hailpern. Observe that Lamport employs a hybrid formalism (TL + Data

Structures), and that in Hailpern's method similar systems, such as FIFO and LIFO, do not have similar specifications. We show that no such enrichment is logically required by introducing an additional axiom within TL which formalizes the assumption that messages accepted by the system can be uniquely identified. In this way, no extraneous formalisms are introduced, and both FIFO and LIFO are expressible with equal ease.

We illustrate our way of specifying message passing systems with temporal logic by three examples (the third example concerns the hierarchical specification of a layered communication network) and draw some conclusions.

This chapter is organized as follows. In section 2 we describe which systems we consider as message passing systems and specialize the requirements of Chapter 2 for these systems in section 3. In section 4 we prove inexpressiveness results for temporal logics with **until** and **since** and their consequence for the specification of message passing systems. Then, in section 5 we review three solutions to overcome the previous logical limitations. We end the chapter with a series of specification examples of message passing systems and draw some conclusions in section 6, respectively section 7.

5.2   What are Message Passing Systems?

In this section we consider message passing systems from the very general and abstract viewpoint of Chapter 2. In particular, message passing systems are viewed as a black box and as long as the observed behavior of two message passing systems is the same as seen from the outside (i.e. in terms of the elements of the interface) they are considered equivalent although the systems may differ internally. A message passing system, then, is a system that gets messages and passes these messages on to their destination. A simple everyday example is a mailbox. The message can be a letter (postcard etcetera) and the message passing system is supplied by the postal company. If we denote the input of a message m by $in(m)$ and the delivery of a message m by $out(m)$, Figure 5.1 represents a message passing system as a black box. So, in and out constitute the abstract interface (see Chapter 2) with the environment and $out(m)$ is considered to be the system reaction on the environment action $in(m)$. Hence, since a message is given by the environment $in(m)$ is the responsibility of the environment and since a message is delivered by the system $out(m)$ is the responsibility of the system. In this

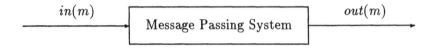

$$in(m) \qquad\qquad out(m)$$

Figure 5.1: Message Passing System as a Black Box

representation the source and destination of a message are left implicit, that
is, $in(m)$ means that there is a source that gives m to the message passing
system and $out(m)$ means that the message passing system delivers m to its
destination (note the asymmetry: the destination of a message must always
be known, while this is not necessarily the case for the source). When sources
and destinations are explicitly represented we get $in(s, m)$ and for symmetry
reasons $out(d, m)$ where, however, always $d = destination(m)$.

The external behavior of a message passing system is characterized by its
input sequence, its output sequence and their relation in time. Hence, only
input, output and their relation determine the observable difference between
several types of message passing systems. This means that quite different
message passing systems such as a simple buffer (or transmission medium)
and a complex communication network should be considered the same as
long as they exhibit the same observable (external) behavior, that is, the
same relation in time between input and output.

The following basic assumption about in, out and their relation in time
is characteristic for all message passing systems:

NC the message passing system does not create messages by itself neither

> **NC1** by creating new messages (a message is new when it has not been
> given to the message passing system before), nor
>
> **NC2** by delivering duplicates of messages given to the message passing
> system.

In other words: the bag of delivered messages is always *some part* of the bag
of messages that have been given to the message passing system. **NC** is an
abbreviation for No Creation. All message passing systems are required to
satisfy this assumption because they are intended to *pass* messages and not to
modify/create or replicate messages. Although it is known that neither **NC1**
nor **NC2** can be guaranteed completely in practice it makes sense to make
such slightly idealized assumptions. Anyway one always has the option of

dropping one or both of them (although in case of dropping **NC1** this would allow the system to exhibit almost any behavior). **NC** is the basic safety assumption for message passing systems in the sense that the system does not commit a bad thing (see e.g. [Lam 83a] for this characterization of safety) by creating messages. Concerning liveness, the basic assumption is that at least some messages that have been given to the system will be delivered at their destination, as formulated in the following liveness assumption:

LA if an infinite number of messages will be given to the message passing
       system, an infinite number of these will be delivered at their destination.

Stated informally, the system may lose an arbitrary number of messages in a row, but eventually it should deliver at least one message (and since time extends to infinity repeating this we get the delivery of a second message, a third message etcetera).

In the above representation of message passing systems we assume that both $in(m)$ and $out(m)$ cause no blocking, that is, the message passing system can never refuse a message that is given to it (it always accepts the message) and it is always able to deliver a message to its destination. In practice this is usually achieved by associating input and output queues at both ends of the message passing system (if we do not make the unrealistic assumption of infinite queues, this implies that $in(m)$ leads to the loss of m when the input queue is full and similarly for $out(m)$ and the output queue).

Because of the physical limitations in the real world it makes sense to make also the following assumption of finite speed:

FS the speed of the message passing system is finite, that is, there is a
       positive (infinite in case the message gets lost) delay between the ac-
       ceptance of a message and its delivery.

As we have seen above, the interface between the message passing system and its environment consists of in and out. Sometimes more information about the interface is available, for example that there is only a single input line or a single output line (a line is called single when at any time there can be at most one message present on the line) leading to the following assumptions no simultaneous input and no simultaneous output:

SI at any moment of time, at most one message can be given to the system,

SO at any moment of time, at most one message can be delivered to its
       destination.

These assumptions apply in particular to the case of a single source and a single destination or in case of explicit representation of sources and destinations for each source and destination separately. Although there cannot be two messages at the same time given to the system nor delivered by the system, it is perfectly possible that there is a message given to the system simultaneously with the delivery of a (different) message by the system. Apart from the assumptions **SI** and **SO** being enforced by the interface it is also possible that the environment, respectively the system, will ensure that no simultaneous inputs, respectively outputs, occur (in spite of the presence of several input, respectively output, lines). This is the reason that the nomenclature single input and single output would be misleading for the above assumptions **SI** and **SO**; therefore we call them no simultaneous input and no simultaneous output, respectively.

In the above description it is not stated whether $in(m)$ and $out(m)$ are considered events (and hence are instantaneous) or actions (and hence have a certain duration). Anyway, for message passing systems it can be assumed that they are events, because it is always possible to identify a unique moment of time at which a message can be said to be accepted, respectively delivered: take for example the case where a message consists of bytes, then one can let $in(m)$ and $out(m)$ correspond to the input (respectively output) of the last byte of m (since we assumed that bytes are not observable but only messages, $in(m)$ can be seen as instantaneous, although on a finer level of granularity the different bytes can be seen).

An example of a message passing system often occurring in practice that is subject to the above restrictions (**NC, LA, FS, SI, SO**) is a transmission medium with a probability between zero and one of a successful transmission. Such a message passing system exhibits only external behaviors that are allowed by these restrictions although the probability of the occurrence of certain behaviors may vary.

Apart from the above restrictions, message passing systems can be distinguished by requiring additional properties. As we saw above the basic liveness requirement for a message passing system is that at least some of the accepted messages will be delivered. Sometimes we need the stronger requirement that *all* accepted messages will eventually be delivered in which case we will call the system perfect. In case messages may get lost (an imperfect system) this notion of a 'lost' message must again be considered as a purely external one, that is, whenever an accepted message is never delivered

it is considered as being lost, although it may remain forever in the message
passing system (and is not lost in the internal view of that system; an example is a network with a routing algorithm that does not guarantee that each
message will eventually reach its destination).

Another distinction can be made by requiring a certain order in which
accepted messages are delivered (if at all). In the above we imposed no order
at all (this corresponds to a bag-like behavior). As an additional requirement
one can pose FIFO ordering (first-in first-out, like queues) or LIFO ordering
(last-in first-out, like stacks). It should be noted, however, that the pure
data structure view of queues and stacks is complicated by the fact that
these can be operated upon in parallel in case of message passing systems by
the input and output of messages (for a stack a simultaneous pop and push,
for example). An example of a FIFO message passing system is an ordinary
buffer. An example of an unordered (i.e., in no order at all) message passing
system is a communication network in which each message is sent on to
an intermediate node depending on some routing algorithm. Due to, for
example, congestion on the chosen route, later messages may arrive earlier
when sent via alternative routes.

5.3   How to Specify Message Passing Systems

Let us review the requirements for a general specification language in Chapter
2 in case of the specification of message passing systems.

Our requirement of syntactical abstractness imposes that the specification
is phrased only in terms of *in*, *out* and messages. A common way to specify
FIFO message passing systems violating this requirement introduces a queue
into the specification and hides it by means of an existential quantifier (see
section 5 of this chapter).

Formal methods for the specification of message passing systems have
been investigated since decades and the results are promising (see for example
[MCS 82],[SM 82]).

Not all such methods conform to the requirement of conformity, for example in section 5 of this chapter we will encounter a method that is well
suited for FIFO message passing systems but awkward for LIFO message
passing systems. Uniformity is also not always guaranteed: a combination of
a logic-based formalism for specifying control and abstract data type theory
for specifying data is in conflict with this requirement. (Note: it may be

that a hybrid formalism can sometimes not be avoided. Nevertheless, when possible a uniform formalism is to be preferred above a hybrid one.)

Because message passing systems are often designed in a layered fashion (with several levels of communication protocols) top-down and bottom-up development are important features of a specification method for such systems.

5.4 Inexpressiveness Results

Our inexpressiveness results concern classes of message passing systems that cannot be characterized in temporal logics with **until** and **since** (see section 3 of Chapter 3). For that purpose we first prove the following special preservation theorem for $L(\textbf{until}, \textbf{since})$.

Definition 5.4.1 Let $\varphi \in L(\textbf{until}, \textbf{since})$, $\mathcal{M}$ be a model, $t \in T$. Define

$$[t]_{\mathcal{M},\varphi} := \{\psi \in SF(\varphi) \mid \mathcal{M}, t \models \psi\}$$

where $SF(\varphi)$ is the set of subformulas of φ (including φ itself).

Definition 5.4.2 Let $\mathcal{M}$ be a model and $t_1, t_2 \in T$ such that $t_1 \leq t_2$. Then $\mathcal{M}_{t_1}^{t_2}$ is the reduction of $\mathcal{M}$ to

$$T_{t_1}^{t_2} := \{t \in T \mid t \leq t_1 \ \vee \ t_2 < t\}.$$

Remark 5.4.3 $\mathcal{M}_{t_1}^{t_2}$ is a submodel of $\mathcal{M}$ but not necessarily a generated submodel of $\mathcal{M}$ (see Definition 3.2.5 in section 2 of Chapter 3).

Theorem 5.4.4 Let $\varphi \in L(\textbf{until}, \textbf{since})$, $\mathcal{M}$ be a model and $t_1, t_2 \in T$ such that $t_1 \leq t_2$ and $[t_1]_{\mathcal{M},\varphi} = [t_2]_{\mathcal{M},\varphi}$. Then for all $t \in T_{t_1}^{t_2}$:

$$\mathcal{M}, t \models \varphi \quad \text{if and only if} \quad \mathcal{M}_{t_1}^{t_2}, t \models \varphi.$$

Proof: By structural induction on φ. We prove the theorem for one of the interesting cases.

Let $\varphi \equiv \varphi_1 \ \textbf{until} \ \varphi_2$, $\mathcal{M}$ be a model and $t_1, t_2 \in T$ such that $t_1 \leq t_2$. Assume

(i) $[t_1]_{\mathcal{M},\varphi} = [t_2]_{\mathcal{M},\varphi}$.

We are going to show that $\mathcal{M}, t \models \varphi$ implies $\mathcal{M}_{t_1}^{t_2}, t \models \varphi$ for $t \leq t_1$. Hence assuming

(ii) $t \leq t_1$ and

(iii) $\mathcal{M}, t \models \varphi_1$ **until** φ_2,

we prove that $\mathcal{M}_{t_1}^{t_2}, t \models \varphi_1$ **until** φ_2.

From (i) and the induction hypothesis we deduce

(iv) $\mathcal{M}, t \models \varphi_1$ implies $\mathcal{M}_{t_1}^{t_2}, t \models \varphi_1$ for all $t \in T_{t_1}^{t_2}$,

(v) $\mathcal{M}, t \models \varphi_2$ implies $\mathcal{M}_{t_1}^{t_2}, t \models \varphi_2$ for all $t \in T_{t_1}^{t_2}$.

From (iii) it follows that

(vi) there exists a $t_0 \in T$ such that $t < t_0$ and $\mathcal{M}, t_0 \models \varphi_2$ and $\mathcal{M}, t' \models \varphi_1$ for all $t' \in T$ such that $t < t'$ and $t' < t_0$.

Distinguish between two cases:

(a) $t_0 \leq t_1$: The result follows in this case immediately from (iv),(v) and (vi)

(b) $t_1 < t_0$: In this case by (ii),(vi) we get also $\mathcal{M}, t_1 \models \varphi_1$ **until** φ_2. By (i) it follows that $\mathcal{M}, t_2 \models \varphi_1$ **until** φ_2. Hence

   (vii) there exists a $t_3 \in T$ such that $t_2 < t_3$ and $\mathcal{M}, t_3 \models \varphi_2$ and $\mathcal{M}, t' \models \varphi_1$ for all $t' \in T$ such that $t_2 < t'$ and $t' < t_3$.

   Because of $t_1 < t_0$ and (vi) we have also

   (viii) $\mathcal{M}, t' \models \varphi_1$ for all $t' \in T$ such that $t < t'$ and $t' \leq t_1$.

   Then $\mathcal{M}_{t_1}^{t_2}, t \models \varphi_1$ **until** φ_2 by (vii) and (viii).

The reverse case $\mathcal{M}_{t_1}^{t_2}, t \models \varphi$ implies $\mathcal{M}, t \models \varphi$ for $t \leq t_1$ can be proved by similar arguments.   ∎

Remark 5.4.5 The result of Sistla et al. (see [SCFG 82]) is obtained by considering only ω-models (see section 3 of Chapter 3) and noting that their operators next-time, until, last-time and since are all expressible in terms of **until** and **since**.

Remark 5.4.6 The theorem can be strengthened to Stavi's language where $\widehat{\text{until}}$ and $\widehat{\text{since}}$ are added, that is, the theorem is also valid for $L(\text{until}, \text{since},$

$\widehat{\text{until, since}}$). We can use similar arguments as in the proof above. To illustrate this we now prove the same case as we treated in the proof above. Let $\varphi \equiv \varphi_1$ **until** φ_2, $\mathcal{M}$ be a model and $t_1, t_2 \in T$ such that $t_1 \leq t_2$ and $[t_1]_{\mathcal{M},\varphi} = [t_2]_{\mathcal{M},\varphi}$. We are going to show that $\mathcal{M}, t \models \varphi$ implies $\mathcal{M}_{t_1}^{t_2}, t \models \varphi$ for $t \leq t_1$. Distinguish between two cases:

(a) $\forall t_3(t < t_3 < t_1 \Rightarrow \mathcal{M}, t_3 \models \varphi_1)$.

Our first aim is to show $\mathcal{M}, t_1 \models \varphi$. In case $t = t_1$ this follows immediately. So suppose $t < t_1$. Since $\mathcal{M}, t \models \varphi_1 \widehat{\text{until}} \varphi_2$, the second conjunct in the definition of **until** (see section 3 of Chapter 3) where t_1 functions as t'' and t_3 as t' leads to $\mathcal{M}, t_1 \models \varphi_1$ and there exists $t_0 > t_1$ such that $\forall t_4(t_1 < t_4 < t_0 \Rightarrow \mathcal{M}, t_4 \models \varphi_1)$. From this we may conclude $\mathcal{M}, t_1 \models \varphi$ as desired. Now, by $[t_1]_{\mathcal{M},\varphi} = [t_2]_{\mathcal{M},\varphi}$ it follows that $\mathcal{M}, t_2 \models \varphi$ and therefore $\mathcal{M}_{t_1}^{t_2}, t \models \varphi$.

(b) There exists a t_3 such that $t < t_3 < t_1$ and $\mathcal{M}, t_3 \models \neg\varphi_1$.

We claim that there also exists a t_4 such that $t < t_4 < t_3$ and $\mathcal{M}, t_4 \models \neg\varphi_1$. Otherwise $\forall t_0 (t < t_0 < t_3 \Rightarrow \mathcal{M}, t_0 \models \varphi_1)$, but then by $\mathcal{M}, t \models \varphi$ it follows that $\mathcal{M}, t_3 \models \varphi_1$, a contradiction. The next claim is that we can find $t_5 \leq t_3$ that fulfills the role of t''' in the third conjunct of the definition of $\widehat{\text{until}}$. Suppose $t_5 > t_3$, then we can conclude $\mathcal{M}, t_3 \models \varphi_2$ because $t < t_3 < t_5$ and $t < t_4 < t_3$ and $\mathcal{M}, t_4 \models \neg\varphi_1$. Now, since $\mathcal{M}, t_3 \models \neg\varphi_1$ and $\mathcal{M}, t_3 \models \varphi_2$ we can as well take $t_5 = t_3$. Since $t_3 < t_1$ this means that all moments involved in the semantics of $\mathcal{M}, t \models \varphi$ precede t_1 so the cut between t_1 and t_2 has no influence upon this. Hence $\mathcal{M}_{t_1}^{t_2}, t \models \varphi$.

We now apply this theorem to prove that many classes of message passing systems cannot be specified in $L(\textbf{until, since})$.

Corollary 5.4.7 The class of all message passing systems (i.e. those systems satisfying the No Creation and basic liveness assumptions **NC** and **LA** of section 2) cannot be specified in $L(\textbf{until, since})$.

Proof: Suppose there exists a formula φ characterizing this class. The number of subformulae of φ is bounded, say by N. Now choose $n > N$ and consider the following model $\mathcal{M}$:

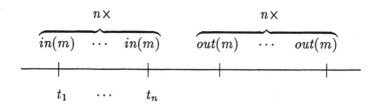

where $m \in Messages$.

This is a possible behavior for this class. Hence φ is satisfied in $\mathcal{M}$. Because $n > N$ there are i, j such that $1 \leq i < j \leq n$ and $[t_i]_{\mathcal{M},\varphi} = [t_j]_{\mathcal{M},\varphi}$. Applying the theorem we conclude that φ is also satisfied in a model with less than n inputs and exactly n outputs. This violates the No Creation assumption. Hence such a φ characterizing this class cannot exist. ∎

Remark 5.4.8 Since the model $\mathcal{M}$ remains a possible behavior when we add any combination of further requirements from section 2 such as finite speed, perfectness and one of the ordering disciplines FIFO and LIFO (since $\mathcal{M}$ uses only one message it is not influenced by such an ordering property) also these classes cannot be specified in $L(\textbf{until}, \textbf{since})$.

Remark 5.4.9 The above proof may not come as a surprise since models like $\mathcal{M}$ represent the context-free language $\{in(m)^n out(m)^n \mid n \in \text{IN}\}$ and propositional temporal logic corresponds to a subset of the ω-regular languages (see e.g. [Tho 86]). However, the above corollary can be strengthened to first-order temporal logic as follows. Because the model $\mathcal{M}$ uses only a finite number of different messages (in this case 1), allowing quantification (using global variables) over the message alphabet (which is here the underlying domain of data) will not help; hence the result can be generalized to this first-order variant with **until** and **since**.

Remark 5.4.10 Since the theorem is also valid for Stavi's language we can strengthen the Corollary and the previous two remarks to the logics where $\widehat{\textbf{until}}$ and $\widehat{\textbf{since}}$ are added.

The essential problem in the specification of message passing systems is that we need both quantification (to account for a possibly infinite message alphabet) and, more importantly, the coupling of a reaction to the unique action that caused this reaction (to account for the counting of an unbounded number of inputs of the same message). Hence, we could not demand that to each $out(m)$ in a row of n there corresponded a unique $in(m)$. To be even more specific, the problem is to specify assumption **NC2** of section 2 forbidding the duplication of messages given to the system. This fact is obvious when inspecting the above proof of the Corollary: the model $\mathcal{M}$ in that proof is clearly involved with the problem of duplication.

5.5 Extensions of Temporal Logic

In this section we consider three solutions to overcome the logical limitations of the previous section.

One possibility is the addition of special data structures to characterize the internal behavior of a system, for example queues for FIFO-behavior, stacks for LIFO-behavior etcetera. In the final specification these special data structures are hidden semantically by means of an existential quantifier. One advocate of this approach is Lamport (see e.g. [Lam 85]). We feel that this approach is not in accordance with several of the desired properties for a specification methodology mentioned in Chapter 2:

1. using an additional internal data structure is implementation biased and as such violates the syntactical abstractness requirement,

2. the behavior of the additional component is described by an additional formalism such as abstract data types, and hence the method loses its uniformity,

3. for different applications one has to plug in different additional components which is in conflict with the conformity requirement.

A second approach is to add special auxiliary variables and operations on them with fixed interpretations. One example of this is history variables with the prefix relation as in the work of Hailpern (see e.g. [Hai 80]). In our opinion, a problem with this approach is that it is biased towards certain behaviors: for specifying FIFO this method is well suited, but awkward for other ordering disciplines such as LIFO. In general one then has to use projections on histories to access the individual elements. What one would like to have is a set of operations on histories such that one can specify each application in terms of this set (such as done for specifying safety properties in

[ZRE 85]). So in this case there is a conflict with the conformity requirement.

Note that in these approaches incoming messages are implicitly made
unique by their place in the data structure, respectively, the history. This
resolves the coupling of a reaction to a unique action. In [KR 85] a third approach can be found in which the unique identification of incoming messages
is explicitly assumed on beforehand, for instance by means of *conceptual* time
stamps. This assumption can be justified by the notion of data-independence
of [Wol 86]. Informally, a system is called data-independent when the values
of the supplied data do not influence the functional behavior of the system.
Since message passing systems are intended to *pass* data, they can be viewed
as being data-independent. One of the results of [Wol 86] implies that the
correctness of a data-independent system does not depend on the uniqueness
of the incoming data. Hence this assumption of unique identification is not
really a restrictive one.

Another look at the assumption of unique identification is provided by
seeing the message passing system as embedded in an additional interface
handling the conceptual time stamps (or counters for that matter) as in Figure 5.2. Here, unique identification transforms an old message m into a pair

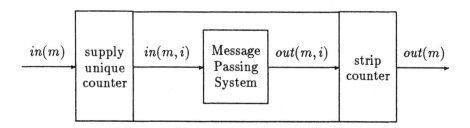

Figure 5.2: Unique Identification by Using Counters

(m, i) where i is a unique identification. As a side remark, this transforma-
tion also gives us the possibility to use unique identification in the case when
a system does not only pass messages but also performs a certain operation,
say f, on them (unique identification might seem problematic in this context
at first sight since f need not be injective). Now an old message m will be
transformed into the pair $(f(m), i)$ whence the input of two messages will
still lead to the output of two different messages despite the fact that f may

transform two different old messages into an identical result.

Although the use of time stamps enforces infinitely many messages even in the case of a finite message alphabet, it is again data-independence that still allows for propositional reasoning: [Wol 86] shows how for a data-independent system properties over an infinite data domain may be reduced to properties over a finite data domain. The advantages of assuming unique identification are threefold:

1. syntactical abstractness: the only predicates are $in(m)$ and $out(m)$,

2. uniformity: the specifications remain purely temporal,

3. conformity: in [KR 85] it is demonstrated that by slight changes of the specification we can describe different properties of systems (e.g. whether it can lose messages or not, whether the ordering is FIFO or LIFO etcetera, see subsection 6.1 of this chapter).

As a consequence of our decision to describe the relation between events in a purely temporal way, the resulting specifications can become rather elaborate. This might be alleviated by modularizing the specification of a system into groups of axioms describing a particular aspect (e.g. subcomponent) of this system.

5.6 Specification Examples

In this section we illustrate the application of temporal logic to message passing systems by a series of examples. The first example treats pure message passing systems, example two is a two-way message passing system with the possibility to close one or both sides of the system and the last example gives a hierarchical specification of a layered communication network.

In our specifications we assume not only linearity of the ordering but also succession towards the future in order to reason about infinite behavior, for instance message passing systems may operate forever. In particular we think of standard models like the natural numbers, the integers, the (non-negative) rational and real numbers.

The priority of operators in the specification examples is as follows: unary operators have the highest priority followed by **until** and **since**-like operators (including the **unless**-operator defined below), then come ∧ (conjunction)

and ∨ (disjunction) and the least priority is given to → (implication) and ↔
(equivalence). With respect to priority, universal and existential quantification are treated as unary operators.

We need several additional temporal operators in our specifications. For
unary temporal operators we showed in section 2 of Chapter 4 how to make
these reflexive. Recall from that section how the reflexive closure of $\mathbf{M}^R$ and
$\mathbf{L}^R$ was defined:

$$\dot{\mathbf{M}}^R \varphi := \varphi \vee \mathbf{M}^R \varphi$$
$$\text{and}$$
$$\dot{\mathbf{L}}^R \varphi := \varphi \wedge \mathbf{L}^R \varphi.$$

In particular we will use $\dot{\mathbf{P}}$, the reflexive version of the $\mathbf{P}$-operator and similarly $\dot{\mathbf{F}}$ and $\dot{\mathbf{G}}$ (for the latter two we will use instead their more usual representation in computer science $\Diamond$, respectively $\Box$, see section 4 of Chapter
3). Apart from these reflexive operators we also need a weak version of the
until denoted by **unless** which does not require that its second argument
will become true eventually:

$$\varphi_1 \text{ unless } \varphi_2 := \mathbf{G} \varphi_1 \vee \varphi_1 \text{ until } \varphi_2.$$

In the specifications we leave out universal quantifications over the data
domains (so all free variables ranging over a data domain should be universally quantified by a series of universal quantifiers in front of the given
axiom).

In the following we only specify the required behavior of the system in its
environment. The specification of the interface can be immediately derived
from the informal description of the embedding of the system in its environment. For example, in case of message passing systems section 2 gives all
relevant information: $in(m)$ is an event with parameter m (an element from
the message domain) for which the environment is responsible and which is
directed from the environment to the system; similarly, $out(m)$ is an event
for which the system is responsible and which is directed from the system to
the environment. When the interface is that simple, a separate specification
becomes superfluous.

The numbering of the axioms of a specification obeys the following conventions. Closely related axioms have the same number ending with a,b
etcetera (e.g. axioms 4a and 4b). ′ denotes replacement of the corresponding
axiom by another (e.g. axiom 3′ replaces axiom 3). Whenever x is added to

the numbering this involves an additional axiom for special cases (e.g. axiom 5x supplements axiom 5).

5.6.1   Example 1: Pure Message Passing Systems

We refer to sections 2 and 5 for the definition of message passing systems and the background on the application of temporal logic to the specification of these systems. Recall from section 2 that *in* and *out* are considered as events (and hence are instantaneous) and that they do not cause blocking. These two features enable us to model *in* and *out* by (unary) predicates.

First we formulate our assumption about the uniqueness of *incoming* messages (the Unique Identification assumption):

$$\textbf{MP1} \qquad in(m) \ \rightarrow \ \neg\, \textbf{D}\, in(m).$$

Here and in the sequel, MP is an abbreviation for message passing. This axiom could be formulated in several equivalent ways such as $\textbf{A} \neg in(m) \ \vee \ \textbf{U}\, in(m)$ or $(in(m) \wedge \textbf{D}\, in(m')) \ \rightarrow \ m' \neq m$, but in any case the most natural way of specifying that $in(m)$ does not occur twice is by using the **D**-operator in some form. Apart from the technical reasons for introducing it in Chapter 4, this gives also an indication for the practical usefulness of this operator. Under this Unique Identification assumption the most important basic assumption of message passing systems, No Creation (see section 2) can be specified by:

$$\textbf{MP2a} \qquad out(m) \ \rightarrow \ \dot{\textbf{P}}\, in(m)$$
$$\textbf{MP2b} \qquad out(m) \ \rightarrow \ \neg\, \textbf{D}\, out(m).$$

The first of these two axioms represents the demand that a message passing system does not create *new* messages while the second axiom represents the absence of duplicate messages (since the input consists of unique messages by the Unique Identification assumption, the output must also consist of unique messages because no messages may be created by the message passing system). Of course these two axioms can be combined into one:

$$\textbf{MP2} \qquad out(m) \ \rightarrow \ \dot{\textbf{P}}\, in(m) \ \wedge \ \neg\, \textbf{D}\, out(m).$$

Notice that the axioms **MP2a** and **MP1** taken together imply that $in(m) \rightarrow \neg\textbf{P}out(m)$ because $\textbf{U}\varphi$ and $\psi \rightarrow \dot{\textbf{P}}\varphi$ imply $\varphi \rightarrow \neg\textbf{P}\psi$.

In general, when perfectness of the message passing system is not assumed, the basic liveness assumption from section 2 is essential to ensure that at least *some* messages arrive (otherwise the system that throws all messages away would satisfy all conditions for a message passing system):

$$\textbf{MP3} \qquad \textbf{G F} \exists\, m\; in(m) \;\rightarrow\; \textbf{F} \exists\, m\; out(m).$$

In section 2 also the assumption of finite speed is mentioned for realistic purposes. Finite speed can be enforced by replacing the $\dot{\textbf{P}}$-operator in axiom **MP2a** above by its strict (i.e. irreflexive) version **P** and similarly for axiom **MP2**:

$$\textbf{MP2a}' \qquad out(m) \;\rightarrow\; \textbf{P}\, in(m)$$
$$\textbf{MP2}' \qquad out(m) \;\rightarrow\; \textbf{P}\, in(m) \;\wedge\; \neg\, \textbf{D}\, out(m).$$

No simultaneous input and no simultaneous output can be specified respectively by

$$\textbf{MP4a} \qquad in(m) \wedge in(m') \;\rightarrow\; m' = m$$
$$\textbf{MP4b} \qquad out(m) \wedge out(m') \;\rightarrow\; m' = m.$$

This concludes the survey of the first set of assumptions for message passing systems. We now turn to the additional assumptions about perfectness and ordering. The perfectness of a message passing system (which implies the basic liveness assumption above) can be expressed by

$$\textbf{MP3}' \qquad in(m) \;\rightarrow\; \Diamond\, out(m).$$

When finite speed is assumed, the $\Diamond$ in the axiom above can be replaced by its strict version **F**. What remains is the specification of special orderings of the output with respect to the input. We look at the cases of FIFO (queue-like) and LIFO (stack-like). First-in first-out requires the same ordering in the output as in the input:

$$\textbf{MP5} \qquad out(m) \wedge \textbf{P}\, out(m') \;\rightarrow\; \dot{\textbf{P}}\,(in(m) \wedge \dot{\textbf{P}}\, in(m')).$$

The above axiom suffices when no simultaneous output is assumed. Otherwise also the case when two messages are output at the same time should be considered. This is reflected in the following axiom:

$$\textbf{MP5x} \qquad out(m) \wedge out(m') \;\rightarrow\; \dot{\textbf{P}}\,(in(m) \wedge in(m')).$$

This exception is caused by the following asymmetry between input and output when requiring FIFO-behavior:

$$in(m')$$

$$in(m) \qquad out(m) \qquad out(m')$$

is allowed (when m and m' are input at the same time none of these messages can be said to have come in first, so they may be output in an arbitrary order), but

$$out(m')$$

$$in(m) \qquad in(m') \qquad out(m)$$

is not (when m is input before m', it should also come out first in the output).

For last-in first-out we get similar specifications, although a bit more complicated because stack-like behavior allows apart from the reversal of the ordering from output and that from input also the possibility that a message has already been output by the system in the meantime, whence a comparison with a message that has been input after that is not needed any more:

MP6 $out(m) \wedge \mathbf{P}\, out(m') \rightarrow$
   $\mathbf{P}\, (in(m') \wedge \dot{\mathbf{P}}\, in(m)) \vee \mathbf{P}\, (out(m') \wedge \neg\, \mathbf{P}\, in(m)).$

Here we consider

$$out(m')$$

$$in(m') \qquad in(m) \qquad out(m)$$

as correct LIFO-behavior (otherwise the last $\mathbf{P}$ in the axiom above should be replaced by its reflexive version $\dot{\mathbf{P}}$). This is comparable with a simultaneous pop and push (recall from section 2 that input and output on both sides of our queues and stacks can operate in parallel, for instance the case $in(m) \wedge out(m')$ is always possible, also when assuming no simultaneous input and no simultaneous output). Just as in the FIFO-case, when no simultaneous output is not assumed, an additional axiom is needed, in this case:

MP6x $out(m) \wedge out(m') \rightarrow$
$$(\neg (in(m) \vee in(m')) \rightarrow \mathbf{P} (in(m) \wedge in(m'))).$$

Again there is a little complication, this time because of the correct LIFO-behavior (unless we suppose finite speed):

(although m' comes in last, m can be considered to have been already output).

In the above account we mixed axioms representing environment assumptions (for example the unique identification assumption) and axioms representing system requirements (for example no creation). A clearer distinction between these two classes of axioms can be provided by writing the specification in the form

$$A_1,\ldots, A_m \Rightarrow A'_1,\ldots, A'_n$$

where $A_1,\ldots, A_m$ are the environment assumptions and $A'_1,\ldots, A'_n$ the system requirements. As an example we give the specification of a perfect, finite speed message passing system with no simultaneous input and no simultaneous output in this form:

$$in(m) \rightarrow \neg \mathbf{D} \, in(m),$$
$$in(m) \wedge in(m') \rightarrow m' = m$$
$$\Rightarrow$$
$$out(m) \rightarrow \mathbf{P} \, in(m) \wedge \neg \mathbf{D} \, out(m),$$
$$in(m) \rightarrow \mathbf{F} \, out(m),$$
$$out(m) \wedge out(m') \rightarrow m' = m.$$

This example made use of PML$(<,>,\neq)$.

5.6.2   Example 2: Channel with Disconnect

In this example we consider a channel between two endpoints 'a' and 'b'. The original informal specification is contained in [DHJR 85]:

The 'channel' between endpoints 'a' and 'b' can pass messages in both directions simultaneously, until it receives a 'disconnect' message from one end, after which it neither delivers nor accepts messages at that end. It continues to deliver and accept messages at the other end until the 'disconnect' message arrives, after which it can do nothing. The order of messages sent in a given direction is preserved.

The channel can be seen as a two-way message passing system as in Figure 5.3. By e we denote one of the endpoints, that is, $e \in \{a, b\}$, and $\tilde{e}$ will denote

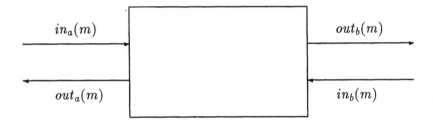

Figure 5.3: Channel with Disconnect

the other endpoint (i.e. $\tilde{a} = b$ and $\tilde{b} = a$). The pairs in_a, out_b and in_b, out_a form a message passing system with FIFO-ordering. Therefore we assume:

the Unique Identification assumption (MP1) for in_e

No Creation and finite speed (MP2') for $in_e, out_{\tilde{e}}$

no simultaneous input and output (MP4a,b) for in_e, out_e

first-in first-out (MP5) for $in_e, out_{\tilde{e}}$.

The only non-standard part of this double message passing system concerns the possibility of a disconnect message. By $disconnect(m)$ we will denote that m is a disconnect message. Input of a disconnect message at one of the two sides causes the closing of that side (for the case of output of a disconnect message, see Remark 5.6.2 below). This can be described by

CD1 $in_e(m) \wedge disconnect(m) \rightarrow \mathbf{G} \left(\neg \exists m \left[in_e(m) \vee out_e(m) \right] \right).$

So, after the input of a disconnect message at e the channel does not accept nor deliver any message any more at that side. The delivery of messages is indeed under control of the channel, but what about the input of messages? In section 2 we gave a representation of message passing systems that allowed no blocking of the input, that is, the system always accepts a message given to it. Also stated there is that this is usually achieved by the association of input and output queues. In normal cases the no blocking assumption makes sense because it abstracts from the subtle difference between the input of a message by the environment and the acceptance of that message by the system. Returning to our example, messages can still be given to a side after the input of a disconnect message but the channel will not accept such messages. In terms of the input queue the message can be put in the queue but the channel will not pass it to the other side.

The remaining property of message passing systems that we did not consider so far is perfectness. In this case the two message passing systems are conditionally perfect, viz. perfect unless disconnected. To describe the state of being disconnected define

$$disconnected_e := \mathbf{P} \,\exists\, m \,[in_e(m) \wedge disconnect(m)].$$

Now, perfect unless disconnected can be specified by

CD2 $in_e(m) \rightarrow \Diamond\,(out_{\bar{e}}(m) \vee disconnected_{\bar{e}}).$

In this axiom we need not additionally assume $\neg\, disconnected_e$ in the antecedent because $in_e(m) \wedge disconnected_e$ cannot occur according to axiom **CD1**.

Remark 5.6.1 Axiom **CD2** allows the channel to delay messages very long and wait for a disconnect message so that no message needs to be delivered. Only if there will be no disconnect at a side, the channel is obliged to deliver the accepted messages eventually.

Remark 5.6.2 In the above a disconnect message is considered as a normal message, but $out_e(m) \wedge disconnect(m)$ does not lead to closing of that side (only input of a disconnect message leads to closing). If also the output of a disconnect message should lead to closing, in axiom **CD1** above the antecedent should be changed into $(in_e(m) \vee out_e(m)) \wedge disconnect(m).$

Remark 5.6.3 When loss of messages is allowed, axiom **CD2** must be replaced by the following conditional liveness requirement:

CD2′ $\mathbf{G}\,\mathbf{F}\,\exists\, m\, in_e(m) \rightarrow \mathbf{F}\,(\exists\, m\, out_{\bar{e}}(m) \vee disconnected_{\bar{e}}).$

Even if the output of a disconnect message leads to closing of that side, the *disconnected*$_{\bar{e}}$ is needed because the disconnect message can get lost (otherwise its arrival at e would lead to closing of endpoint e and the premiss **G F** $\exists\, m\, in_e(m)$ could never be fulfilled).

In this example we made use of $PML(<,>,\neq)$.

5.6.3   Example 3: Layered Communication Network

Introduction

In this example we consider a communication network consisting of three levels and layers, see Figure 5.4. Although we are aware that the usual num-

		level 1
layer 1 (end-to-end)	$in(n,m)$, $out(n,m)$	
		level 2
layer 2 (packets)	$in(n,p)$, $out(n,p)$	
		level 3
layer 3 (intermediate nodes)	$transmit(p,n,i)$, $arrive(p,n,i)$	

Figure 5.4: Layered Communication Network

bering for layered networks is the other way around (lowest layer is numbered 1 as in the ISO OSI model), the given numbering is the most convenient for the current example. On level 1 there are messages and nodes and the service provided by layer 1 is end-to-end reliable message passing using $in(n,m)$ (node n sends message m) and $out(n,m)$ (m is delivered at its destination node n). This is a perfect message passing system with multiple sources and destinations as treated in section 2 of this chapter. As is also given there, the relation between the delivery of a message and the destination of that message is given by

$$out(n,m) \;\rightarrow\; n = destination(m).$$

On the second level the messages are decomposed into packets and the service provided by layer 2 is end-to-end reliable packet passing using $in(n,p)$ (node n sends packet p) and $out(n,p)$ (p is delivered at its destination node

n). This is a perfect packet passing system with multiple sources and destinations. In general, the difference between a message and a packet is that a
packet usually has a fixed size while the length of a message can be arbitrary
(and often even unbounded). When all packets of a message have arrived at
the destination the message will be delivered. For the delivery of a packet
and the destination of that packet the same relation holds as for messages
above:

$$out(n, p) \;\rightarrow\; n = destination(p).$$

The relation between a message and the packets into which it is decomposed
is as follows. This relation is characteristic (giving the minimal demands) for
message segmenting protocols. By $p \in m$ we denote that p is amongst the
packets into which m is decomposed. Each message consists of at least one
packet:

$$\exists p \;\; p \in m.$$

On the other hand, a message is decomposed only in a finite number of
packets. Therefore, instead of $\forall p \; [p \in m \;\rightarrow\; ...]$ we will henceforth write
$\bigwedge_{p \in m} ...$ and similarly $\bigvee_{p \in m} ...$ instead of $\exists p \; [p \in m \land ...]$. In order to be
able to decide at the destination of a packet to which message it belongs we
assume that each packet belongs to at most one message:

$$p \in m \land p \in m' \;\rightarrow\; m' = m.$$

Furthermore, the destination of a packet which belongs to a message must
obviously be the same as the destination of that message:

$$p \in m \;\rightarrow\; destination(p) = destination(m).$$

On level 3 a network of intermediate nodes is introduced via which packets
are transmitted towards their destination. The service provided by layer
3 is point-to-point reliable transmission using $transmit(p, n, i)$ (packet p is
transmitted from node n to node i) and $arrive(p, n, i)$ (packet p coming from
node n arrives at node i). The transmission medium between two such nodes
n and i provides a perfect packet passing system. A packet traveling on the
way to its destination may traverse an intermediate node more than once:
sometimes a packet can come back, for example because an intermediate
node in the network decides to reroute the packet (and incidentally the new
route traverses old intermediate nodes) due to congestion of the network in
a certain direction. This entails a complication for the unique identification
assumption about packets at this level.

Layer 3 assumes the availability of perfect transmission media. The next layer in this hierarchical communication network could be the implementation of such perfect transmission media by means of imperfect ones using acknowledgments and time-out for retransmission. Since such a layer involves quantitative temporal properties the specification of such a fourth layer belongs to the next chapter. In example 9 of section 5 of that chapter we will specify such an implementation of reliable communication by means of time-out and retransmission over imperfect transmission media.

Layer 1

This layer provides a perfect message passing system with multiple sources and destinations. Because there are multiple sources and destinations the formulation of the unique identification assumption about messages must also take into account messages that originate from different sources as is done in the following two axioms:

$$in(n, m) \; \rightarrow \; \neg \, \mathbf{D} \, in(n', m)$$

$$in(n, m) \wedge in(n', m) \; \rightarrow \; n' = n.$$

The last axiom could be viewed as the opposite of the no simultaneous input assumption (see Example 1): two different sources (nodes n and n') may not generate the same message at the same time (for different moments in time this is ensured by the first axiom). In practice, this is normally anyway the case because a message usually includes a field for the source of the message.

In order not to have to deal with the exceptional case of the input of a message at its destination in the sequel, we assume for ease of presentation that this will not happen:

$$\neg \, in(destination(m), m).$$

As we have seen in Example 1, the assumptions of No Creation and finite speed can be taken together, in the case of multiple sources and destinations as follows:

$$out(n, m) \; \rightarrow \; \mathbf{P} \, \exists \, n' \, in(n', m) \wedge \neg \, \mathbf{D} \, out(n, m).$$

Remember from the Introduction of this example that out obeys the requirement

$$out(n, m) \; \rightarrow \; n = destination(m).$$

The only remaining property left is perfectness:

$$in(n,m) \rightarrow \Diamond\, out(destination(m), m).$$

Layer 2

This layer provides a perfect packet passing system with multiple sources and destinations. The only difference with layer 1 is the sort of data that is passed: packets instead of messages. The following list of axioms is derived from that of layer 1 by substituting the packet variable p for the message variable m:

$$in(n,p) \rightarrow \neg\, \mathbf{D}\; in(n',p)$$

$$in(n,p) \wedge in(n',p) \rightarrow n' = n$$

$$\neg\, in(destination(p), p)$$

$$out(n,p) \rightarrow \mathbf{P}\, \exists\, n'\; in(n',p) \wedge \neg\, \mathbf{D}\; out(n,p)$$

$$out(n,p) \rightarrow n = destination(p)$$

$$in(n,p) \rightarrow \Diamond\, out(destination(p), p).$$

Relating Layer 1 and Layer 2

As we described in the Introduction of this example, the second level first disassembles a message into packets, sends the packets through the packet passing system provided by layer 2 and finally reassembles the packets into the message at the destination. So, pictorially layer 1 can be represented as in Figure 5.5. Recall from the Introduction of this example the relation between packets and messages obeying the following axioms:

$$\exists\, p\;\; p \in m$$

$$p \in m \wedge p \in m' \rightarrow m' = m$$

$$p \in m \rightarrow destination(p) = destination(m).$$

Furthermore, we write $\bigwedge_{p \in m} \ldots$ instead of $\forall\, p\, [p \in m \rightarrow \ldots]$ and $\bigvee_{p \in m} \ldots$ instead of $\exists\, p\, [p \in m \wedge \ldots]$ because a message can only be disassembled into a finite number of packets.

In order to describe the relation between layer 1 and layer 2 we have to specify the connection between $in(n,m)$ and $in(n,p)$ via the disassembling

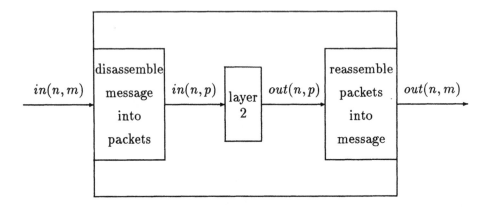

Figure 5.5: Representation of Layer 1

of messages, respectively the connection between $out(n,p)$ and $out(n,m)$ via the reassembling of packets.

First, the input of a message at a node leads to the sending of all its packets from that node into the packet passing system of layer 2:

$$in(n,m) \quad \rightarrow \quad \bigwedge_{p \in m} \Diamond \, (in(n,p) \wedge \neg\, \mathbf{D} \, in(n,p)).$$

The part $\neg\, \mathbf{D} \, in(n,p)$ ensures that a packet is sent only once and relies on the unique identification of messages. Reversely, a packet may only be sent from a node into the packet passing system of layer 2 when it is part of a message that has been input at that node before:

$$in(n,p) \quad \rightarrow \quad \exists \, m \, [\, p \in m \wedge \dot{\mathbf{P}} \, in(n,m)\,].$$

At the other side, the arrival of all packets that constitute a message leads to the output of that message:

$$\bigwedge_{p \in m} \dot{\mathbf{P}} \, out(n,p) \wedge \bigvee_{p \in m} out(n,p) \quad \rightarrow \quad \Diamond \, out(n,m).$$

Reversely, a message may only be output when all its packets have arrived and it has not been output before (in order to avoid duplication of messages):

$$out(n,m) \quad \rightarrow \quad \neg\, \mathbf{P} \, out(n,m) \wedge \bigwedge_{p \in m} \dot{\mathbf{P}} \, out(n,p).$$

These four axioms describe precisely the relationship between $in(n, m)$ and $in(n, p)$, respectively $out(n, p)$ and $out(n, m)$. Having defined these relationships we can ask ourselves whether the second level is a correct refinement of the first level, that is, whether we can prove from the specification of layer 2 and the above relationship between layer 1 and layer 2 that the specification of layer 1 is fulfilled. To this end we have to prove all axioms of layer 1 except of course the assumptions of layer 1 about its environment, namely the two axioms about the unique identification of messages and the axiom about not inputting a message at its destination.

No creation of new messages is proved as follows.

> Suppose $out(n, m)$. Then by the relation between $out(n, m)$ and $out(n, p)$ it follows that $\bigwedge_{p \in m} \mathbf{P}\, out(n, p)$. Since $\exists p\, p \in m$ (the first axiom relating packets and messages) this implies $\bigvee_{p \in m} \dot{\mathbf{P}}\, out(n, p)$. The no creation of new packets axiom for layer 2 gives us $\bigvee_{p \in m} \mathbf{P} \exists n' in(n', p)$. The relation between $in(n, p)$ and $in(n, m)$ then implies $\bigvee_{p \in m} \exists n'\, \mathbf{P} \exists m'\, [p \in m' \wedge \dot{\mathbf{P}}\, in(n', m')]$. Now, the second axiom relating packets and messages ($p \in m \wedge p \in m' \rightarrow m' = m$) gives $m' = m$, so we may conclude $\bigvee_{p \in m} \exists n'\, \mathbf{P}\, \dot{\mathbf{P}}\, in(n', m)$. By leaving out p (which plays no role any more) and contracting the $\mathbf{P}$ and $\dot{\mathbf{P}}$ we arrive at the desired conclusion $\mathbf{P} \exists n'\, in(n', m)$.

The second part of the No Creation requirement, no duplication of messages is easier: from the last axiom of the four axioms relating layer 1 and layer 2 it follows that $out(n, m) \rightarrow \neg \mathbf{P}\, out(n, m)$ and hence $out(n, m) \rightarrow \neg \mathbf{D}\, out(n, m)$ (since $\mathbf{D}\varphi \equiv \mathbf{P}\varphi \vee \mathbf{F}\varphi$ for linear orderings).

Next we have to show that $out(n, m) \rightarrow n = destination(m)$. As above we can derive from $out(n, m)$ that $\bigvee_{p \in m} \dot{\mathbf{P}}\, out(n, p)$. The axiom $out(n, p) \rightarrow n = destination(p)$ of layer 2 then implies that $\bigvee_{p \in m} n = destination(p)$. By the third axiom relating packets and messages ($p \in m \rightarrow destination(p) = destination(m)$) the desired conclusion $n = destination(m)$ follows.

The final axiom of layer 1 to be proved is perfectness:

$$in(n, m) \rightarrow \Diamond\, out(destination(m), m).$$

We prove this as follows.

> Suppose $in(n, m)$. By the relation between $in(n, m)$ and $in(n, p)$ this implies $\bigwedge_{p \in m} \Diamond\, in(n, p)$. By the perfectness of layer 2 we get

$\bigwedge_{p \in m} \Diamond \Diamond out(destination(p), p)$. Contracting $\Diamond \Diamond$ into a single $\Diamond$ and noting that the finite conjunction leads to a moment when all packets of m have reached their destination we may conclude from this $\Diamond (\bigwedge_{p \in m} \dot{\mathbf{P}} out(destination(p), p) \wedge \bigvee_{p \in m} out(destination(p), p))$. By the third axiom relating packets and messages ($p \in m \rightarrow destination(p) = destination(m)$) this transforms into $\Diamond (\bigwedge_{p \in m} \dot{\mathbf{P}} out(destination(m), p) \wedge \bigvee_{p \in m} out(destination(m), p))$. Now, by the relation between $out(n, p)$ and $out(n, m)$ this implies $\Diamond \Diamond out(destination(m), m)$ so that again contracting $\Diamond \Diamond$ into $\Diamond$ yields the desired conclusion.

The above proof is given on a semantical level. As an illustration we show how such an argument can be transformed into a formal proof:

1. $in(n, m) \rightarrow \bigwedge_{p \in m} \Diamond (in(n, p) \wedge \neg \mathbf{D}\, in(n, p))$
   (relation $in(n, m)$ and $in(n, p)$)

2. $\Diamond (\varphi \wedge \psi) \rightarrow \Diamond \varphi$ (temporal logic)

3. $in(n, m) \rightarrow \bigwedge_{p \in m} \Diamond in(n, p)$ (1,2)

4. $in(n, p) \rightarrow \Diamond out(destination(p), p)$ (perfectness layer 2)

5. $in(n, m) \rightarrow \bigwedge_{p \in m} \Diamond \Diamond out(destination(p), p)$ (3,4)

6. $\Diamond \Diamond \varphi \rightarrow \Diamond \varphi$ (temporal logic over linear frames)

7. $in(n, m) \rightarrow \bigwedge_{p \in m} \Diamond out(destination(p), p)$ (5,6)

8. $\Diamond \varphi_1 \wedge \Diamond \varphi_2 \rightarrow \Diamond ((\varphi_1 \wedge \dot{\mathbf{P}} \varphi_2) \vee (\varphi_2 \wedge \dot{\mathbf{P}} \varphi_1))$
   (temporal logic over linear frames)

9. $\bigwedge_{i=1}^{n} \Diamond \varphi_n \rightarrow \Diamond (\bigwedge_{i=1}^{n} \dot{\mathbf{P}} \varphi_n \wedge \bigvee_{i=1}^{n} \varphi_n)$ (repetition of 8)

10. $in(n, m) \rightarrow$ (7,9)
   $\Diamond (\bigwedge_{p \in m} \dot{\mathbf{P}} out(destination(p), p) \wedge \bigvee_{p \in m} out(destination(p), p))$

11. $p \in m \rightarrow destination(p) = destination(m)$
    (third axiom relating packets and messages)

12. $in(n, m) \rightarrow$ (10,11)
    $\Diamond (\bigwedge_{p \in m} \dot{\mathbf{P}} out(destination(m), p) \wedge \bigvee_{p \in m} out(destination(m), p))$

13. $\bigwedge_{p\in m} \dot{\mathbf{P}}\, out(n,p) \wedge \bigvee_{p\in m} out(n,p) \;\rightarrow\; \Diamond\, out(n,m)$
    (relation $out(n,p)$ and $out(n,m)$)

14. $in(n,m) \;\rightarrow\; \Diamond\Diamond\, out(destination(m),m)$ $\hfill$ (12,13)

15. $in(n,m) \;\rightarrow\; \Diamond\, out(destination(m),m)$ $\hfill$ (6,14)

Notice that we used twice in this proof that we are working over linear frames: we used transitivity in 6 and comparability in 8.

Having proved that all axioms of layer 1 except its environment assumptions are satisfied is not yet sufficient to prove that the first level has been correctly refined. We also have to show that the environment assumptions made by layer 2 are met since the second level should take care of that.

Firstly, suppose that $in(n,p) \wedge \mathbf{D}\, in(n',p)$. We have to show that this leads to a contradiction. By the relation between $in(n,p)$ and $in(n,m)$ and the second axiom relating packets and messages $(p \in m \wedge p \in m' \rightarrow m' = m)$ this assumption leads to $\exists\, m[p \in m \wedge \dot{\mathbf{P}}\, in(n,m) \wedge \mathbf{D}\,\dot{\mathbf{P}}\, in(n',m)]$. By the unique identification assumption about messages of layer 1 it follows that $n' = n$, but then the initial supposition transforms into $in(n,p) \wedge \mathbf{D}\, in(n,p)$. This, however, is impossible because of the relation between $in(n,m)$ and $in(n,p)$: $in(n,m)$ implies $\Diamond\, (in(n,p) \wedge \neg\, \mathbf{D}\, in(n,p))$.

Secondly, suppose $in(n,p) \wedge in(n',p)$. We have to show that $n' = n$. As above this assumption leads to $\exists\, m[p \in m \wedge \dot{\mathbf{P}}\, in(n,m) \wedge \dot{\mathbf{P}}\, in(n',m)]$. Then indeed $n' = n$ by the unique identification assumption about messages of layer 1.

Thirdly and finally we have to show that $\neg\, in(destination(p),p)$. So suppose $in(n,p)$. By the relation between $in(n,p)$ and $in(n,m)$ it follows that $\exists\, m[p \in m \wedge \dot{\mathbf{P}}\, in(n,m)]$. By the axiom about not inputting a message at its destination of layer 1 we may conclude $\exists m[p \in m \wedge n \neq destination(m)]$. By the third axiom relating packets and messages $(p \in m \rightarrow destination(p) = destination(m))$ we reach the desired conclusion $n \neq destination(p)$.

Layer 3 and its relation to Layer 2

On this layer the perfect packet passing system of layer 2 is implemented by a network of nodes through which the packets are sent to their destination. This layer relies on a reliable transmission layer between each pair of (adjacent) nodes and furthermore includes a routing algorithm at each node to determine where incoming packets should go next. Pictorially layer 2 can then be represented as in Figure 5.6.

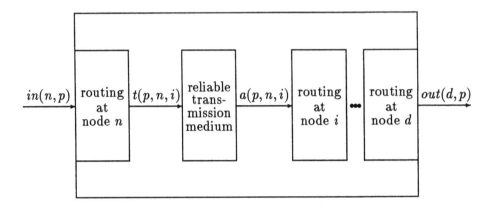

$$t(p, n, i) = transmit(p, n, i), \ a(p, n, i) = arrive(p, n, i), \ d = destination(p).$$

Figure 5.6: Representation of Layer 2

As we described in the Introduction of this example, a packet traveling on the way to its destination may traverse the same intermediate node more than once: a packet can return at a node because of a rerouting decision at another node. This implies that we cannot take the ordinary unique identification assumption for *transmit* in this case. Before we go into the specification of the routing algorithm at the nodes and the reliable transmission medium between a pair of nodes we can at least restrict the way *transmit* handles a packet globally in the network, that is, in relation to different nodes. In particular, at each moment a packet can only be in one place:

$$transmit(p, n, i) \ \wedge \ transmit(p, n', i') \ \rightarrow \ n' = n \ \wedge \ i' = i$$

$$transmit(p, n, i) \ \rightarrow \ \neg \, \exists \, n' \, \exists \, i' \, transmit(p, n', i') \ \textbf{unless} \ arrive(p, n, i).$$

The first axiom ensures this for the moment transmission starts while the second axiom ensures it during transmission (the next transmission can only occur after the transmission layer has delivered the packet).

Now we are going to look at the routing algorithm inside a node. First of all, it should transmit a packet that arrived and for which this node is an intermediate node to a chosen next node:

$$arrive(p, n, i) \ \wedge \ i \neq destination(p) \ \rightarrow \ \textbf{F} \, \exists \, i' \, transmit(p, i, i').$$

Secondly, it may only transmit a packet when that packet arrived at this node (or was input directly from above by *in*) and it may choose a next node only once:

$$transmit(p, i, i') \;\rightarrow$$
$$\neg \exists\, i'\; transmit(p, i, i')\; \textbf{since}\; (in(i, p) \;\vee\; \exists\, n\; arrive(p, n, i)).$$

The first axiom guarantees that a packet will be sent on to the next node but it does not guarantee that the packet will eventually reach its destination. This is, however, not a local property (i.e. a property for a single node) but a global property which must be ensured by the routing algorithms in all nodes taken together. That a packet reaches its destination furthermore depends obviously on the reliability of the transmission media used between intermediate nodes. If these can be assumed to be perfect (which is indeed the case) the collection of routing algorithms guarantees the arrival of a packet at its destination as follows:

$$transmit(p, n, i)\;\wedge\;\square\,(transmit(p, n', i')\;\rightarrow\;\textbf{F}\;arrive(p, n', i'))$$
$$\rightarrow\;\textbf{F}\,\exists\, n'\; arrive(p, n', destination(p)).$$

This requirement is characteristic for routing protocols.

Next we specify the reliable transmission media between pairs of nodes. Since the pair of nodes is fixed for each transmission medium we write simply $transmit(p)$ and $arrive(p)$ instead of $transmit(p, n, i)$, respectively $arrive(p, n, i)$. All the transmission media are perfect packet passing systems. As we remarked already we cannot use the ordinary unique identification assumption for $transmit$ in this case because we allow packets to return at an intermediate node. Hence $transmit(p)\;\wedge\;\textbf{D}\,transmit(p)$ is possible. However, as the new environment assumption we can at least demand that the environment can only provide the same packet for the next time when the previous one has arrived:

$$transmit(p)\;\rightarrow\;\neg\,transmit(p)\;\textbf{unless}\;arrive(p).$$

So, between two transmittals of the same packet there is at least one arrival of that packet. Under this environment assumption the No Creation assumption (together with finite speed) is formulated as follows:

$$arrive(p)\;\rightarrow\;\neg\,arrive(p)\;\textbf{since}\;transmit(p).$$

This implies $arrive(p)\;\rightarrow\;\textbf{P}\,transmit(p)$ taking care that no new packets are created. The other part of No Creation, no duplication of packets is also

taken care of since the above axiom prohibits the possibility of two arrivals
of the same packet after only one transmit of that packet. Perfectness can
be formulated as usual:

$$transmit(p) \rightarrow \mathbf{F}\, arrive(p).$$

We can take the environment assumption and this axiom together (blurring
the distinction between assumptions about the environment and requirements
for the system) to get

$$transmit(p) \rightarrow \neg\, transmit(p)\ \mathbf{until}\ arrive(p).$$

So, the environment waits to transmit a packet again until this packet has
arrived and that arrival is indeed guaranteed. We need the irreflexive operator $\mathbf{F}$ instead of $\diamond$ in the axiom for perfectness in this case to exclude the
following illegal behavior:

Taking also the axiom for the No Creation assumption into account it follows
that, when restricting attention to a single packet, $transmit$ and $arrive$ may
happen only alternatingly starting with $transmit$ and ending with $arrive$
(where the next $transmit$ together with the previous $arrive$ is allowed,
though).

After having specified the entities on this level (the routing algorithms,
the transmission media and their global connection) we are ready to describe
the relation between layer 2 and layer 3. This is done by specifying the
connection between in and $transmit$, respectively $arrive$ and out.

First, the input of a packet at a node leads to the transmittal of this
packet to a chosen node (remember that we assumed that a packet is not
input at its destination):

$$in(n, p) \rightarrow \mathbf{F} \exists\, i\ transmit(p, n, i).$$

Reversely, each $transmit$ must have its root with in:

$$transmit(p, i, i') \rightarrow \mathbf{P} \exists\, n\ in(n, p).$$

At the other side, arrival of a packet at its destination leads to output of that packet:

$$arrive(p, i, destination(p)) \; \rightarrow \; \mathbf{F}\, out(destination(p), p).$$

Reversely, a packet may only be output when it arrived at its destination and it has not been output before (in order to avoid duplication of packets):

$$out(n, p) \; \rightarrow \; \neg\, \mathbf{P}\, out(n, p) \wedge n = destination(p) \wedge \mathbf{P}\, \exists\, i\; arrive(p, i, n).$$

Having defined the relationship between layer 2 and layer 3 by these four axioms we can ask whether the third level is a correct refinement of the second level. To this end we have to prove the axioms (except the environment assumptions) of layer 2 and the environment assumption of layer 3.

No creation of new packets is proved as follows. Suppose $out(n, p)$. By the relation between out and $arrive$ it follows that $\mathbf{P}\,\exists\,i\; arrive(p, i, n)$. The No Creation assumption for the transmission medium between i and n leads now to $\mathbf{P}\,\exists\,i\; transmit(p, i, n)$. The relation between $transmit$ and in then gives the desired conclusion $\mathbf{P}\,\exists\,n'\; in(n', p)$.

The second part of the No Creation assumption (no duplication of packets) is even easier: the last of the four axioms describing the relationship between layer 2 and layer 3 implies that $out(n, p) \;\rightarrow\; \neg\,\mathbf{P}\, out(n, p)$, hence $out(n, p) \;\rightarrow\; \neg\,\mathbf{D}\, out(n, p)$.

The axiom $out(n, p) \;\rightarrow\; n = destination(p)$ of layer 2 follows directly from the last axiom describing the relationship between layer 2 and layer 3.

Perfectness is proved as follows. Suppose $in(n, p)$. By the relation between in and $transmit$ it follows that $\mathbf{F}\,\exists\,i\; transmit(p, n, i)$. The global requirement on the collection of routing algorithms together with the perfectness of the transmission media guarantee that p will arrive at its destination: $\mathbf{F}\,\exists\,n'\; arrive(p, n', destination(p))$. By the relation between $arrive$ and out this leads to $\mathbf{F}\, out(destination(p), p)$, so certainly $\Diamond\, out(destination(p), p)$.

The only remaining axiom to be checked is the environment assumption of layer 3, namely $transmit(p) \;\rightarrow\; \neg\, transmit(p)$ **unless** $arrive(p)$. This follows directly from the stronger restriction

$$transmit(p, n, i) \;\rightarrow\; \neg\,\exists\,n'\,\exists\,i'\; transmit(p, n', i') \; \textbf{unless}\; arrive(p, n, i)$$

for the way the network globally handles the transmittal of packets (at each moment the packet can be in transmission only in one place).

By transitivity of the refinement relation we may also conclude that the third level is a correct refinement of the first level.

This example made use of $L(\mathbf{until}, \mathbf{since})$.

5.7 Conclusions

In this chapter we proved several limitations of temporal logics for the specification of message passing systems. The counterexamples indicate that a necessary ingredient for such a specification is the ability to trace back (in time) every delivered message to its unique moment of acceptance. With this in mind one can take one of two directions. Either one argues that, because it is not expressive enough, temporal logic should be enriched with an additional formalism for reasoning about such systems, *or*, having identified the trouble spot, one makes some general assumptions about these systems that are strong enough to enable a purely temporal specification. The first course is taken by most researchers in the field. This might be caused by lack of recognition of the essential missing ingredients. The second course is attractive since the general assumption about message passing systems, viz. that incoming messages can be uniquely identified, can be translated into the logic and hence can be reasoned with inside the formalism itself.

We illustrated our approach to the specification and verification of message passing systems by three examples. The first example showed how pure message passing systems can still be specified (notwithstanding the inexpressiveness results of section 4) with the classical temporal logic treated in section 2 of Chapter 3 (using only the temporal operators $\mathbf{F}$ and $\mathbf{P}$ since $\mathbf{D}$ is equivalent to the disjunction of these two operators when working over temporal frames with a linear ordering) in an elegant and easy way. The price we had to pay, the unique identification assumption on the incoming data, was shown to be less high than might have been thought at first glance. The second example illustrated that complications such as a two-way message passing system with possibilities to close either side can also be handled quite easily. In fact, the majority of message passing properties of this example could be derived directly and in a straightforward way from the pure message passing properties of Example 1 so that the specification only had to concentrate on non-standard features such as the treatment of the special disconnect message. This suggests that the standard part of our specifications can be 'modularized' in the sense that we can use certain sets of axioms (such as those for a perfect FIFO message passing system) as parts that can be added to a specification maintaining the same restrictions on the required behavior as when imposed in isolation. In the third example we considered a system that was decomposed into subsystems. We showed how such a system can be specified in a hierarchical fashion and how the correctness of the refinement steps can be proved.

Chapter 6

Time-Critical Systems

6.1 Introduction

This chapter is motivated by the need for a formal specification method for time-critical systems. The need for such a method is becoming acute since more and more vital applications such as nuclear power stations, computer controlled chemical plants, flight control software for airplanes, etcetera, are of a time-critical nature. Time-critical systems are characterized by *quantitative* timing properties relating occurrences of events. Typical examples are:

1. Maximal distance between an event and its reaction, for example, every A is followed by a B within 3 time units (a typical promptness requirement).

2. Exact distance between events, for example, every A is followed by a B in exactly 7 time units (as with the setting of a timer and its time-out).

3. Minimal distance between events, for example, two consecutive A's are at least 5 time units apart (assumption about the rate of input from the environment).

4. Periodicity, for example, event E occurs regularly with a period of 4 time units.

5. Bounded response time, for example, there is a maximal number of time units so that each occurrence of an event E is responded to within this bound.

After the development of a characterization for time-critical systems we look at requirements for specification languages in the context of such systems.

Like we did for message passing systems we investigate the possibilities of temporal logic for specifying time-critical systems. Because they only involve qualitative temporal operators it is obvious that the standard temporal logics of Chapter 3 cannot deal with quantitative temporal requirements. Therefore, we extend the usual temporal frames by including a distance function to measure time and analyze what restrictions should be imposed on such a function. This distance function maps two points in time to a value in a metric domain on which addition and a zero are defined. The specification method we propose, called *metric temporal logic* (MTL for short), is based on the polymodal logics of Chapter 4: our metric operators are obtained by indexing polymodal operators by parameters taken from the metric domain. Our philosophy is to extend the pure qualitative view of time of standard temporal logics in a faithful manner in order to reason also about qualitative properties in a convenient way. We succeed in doing this by including also the precedence relation between points in time and showing how the metric parameters of operators can be 'quantified away' to obtain the corresponding qualitative versions. We show how the five quantitative timing properties above can be expressed in metric temporal logic. Concerning qualitative properties, the whole first-order language of linear order can be expressed in metric temporal logic. We also look at the issue of axiomatization.

We illustrate metric temporal logic by means of nine examples involving time-critical (and often also message passing) features amongst which are common real-time constructs such as a time-out and the wait/delay statement of some concurrent programming languages.

This chapter is organized as follows. In section 2 we describe the characteristics of time-critical systems and specialize the requirements of Chapter 2 for these systems in section 3. Section 4 introduces metric temporal logic which is illustrated by means of a series of specification examples of time-critical systems in section 5. At last we present some conclusions in section 6.

6.2 What are Time-Critical Systems?

The most important characteristic of a time-critical system is the demand to keep abreast with an autonomous environment by reacting properly and timely to events occurring in the environment asynchronously from the operation of the system. Therefore, the environment-system interaction (the reaction of the system on the external stimuli from the environment giving rise to a so-called stimulus-response mechanism) is subject to quantitative temporal requirements. These temporal requirements state a relation between occurrences of events and can be classified as follows:

- response time: this relates the timing of the occurrence of an event and its response. The most usual cases are

  ⋆ maximal distance between an event and its response (e.g. time-out)

  ⋆ exact distance (e.g. delay)

- frequency: this relates occurrences of the same event. The most usual cases are

  ⋆ minimal distance between two occurrences (assumption about the rate of stimuli from the environment)

  ⋆ exact distance, also called periodicity (e.g. clocks and samplers).

The first four of the five examples in section 1 correspond directly to the classification above (examples 1 and 2 concern maximal respectively exact response time and examples 3 and 4 concern minimal respectively exact frequency). All these temporal requirements have a quantitative nature and the quantitative elements involved are constants expressed as a certain number of time units. The fifth example in section 1 is in fact the quantified equivalence of the first example. The other examples 2, 3 and 4 have also quantified equivalents, but example 5 is the most common one. The quantitative nature of these temporal requirements is typical for time-critical systems (qualitative temporal requirements occur already in any concurrent system, think of fairness, and even sequential systems, e.g. termination).

Another classification of quantitative temporal requirements relates to the distinction between relative and absolute temporal requirements. Absolute temporal requirements calibrate all occurrences of events to a fixed reference point (the start of the system or the first occurrence of a particular event) while relative temporal properties have no fixed reference point but depend on occurrences of events. In the above four cases periodicity is an absolute temporal requirement (e.g. all later samples can be related to the first sample by means of the sample rate), the other three being relative (the occurrence of an event triggers its response, so the timing of that response can only be related to that occurrence of the event). As will be clear from the above, events play a very important role in time-critical systems.

Since quantitative temporal requirements state a relation between an event in the environment and an event in the system (or between events in different components of a system), these requirements necessarily refer to a global notion of time. This global notion of time should not be identified with the introduction of a global clock: the difference between time and real clocks is that clocks always drift (in other words: time can be considered as a perfect, idealized clock).

In the theory of concurrency the interleaving model plays an important role. One can question the suitability of this model in case real-time and concurrency are combined. Modeling parallel computation by interleaving is a sufficient idealization if only qualitative temporal requirements are involved. As soon as quantitative temporal requirements come into play, however, as in the case of time-critical systems, such an execution model is usually not adequate any more. For example, ensuring maximal distance between events is impossible if some processes can take an arbitrary number of steps while other processes are inactive. In such a case either all processes have their own processor (the maximal parallelism model as in [KSRGA 85]) or some processes share one processor and they are scheduled in such a way that each process gets its turn within bounded time. Furthermore, in some applications data can appear at different places in a truly concurrent way. With respect to the temporal requirements above an arbitrary sequentialization is not appropriate any more. Even stronger, it becomes more and more practice today to incorporate local (co)processors with dedicated tasks (e.g. sampling) into the system so that truly parallel computation is the only realistic model in such a distributed configuration. Notwithstanding the above conceptual objections, [Ost 89] and [HMP 91] show how by a careful incorporation of time into the interleaving model (using lower and upper time bounds on transitions) most timing properties of time-critical systems can be represented.

The most prominent examples of time-critical systems are real-time systems. Real-time systems have additional aspects, however: they not only deal with (quantitative) temporal requirements, but also performance, safety and reliability are essential aspects. Nevertheless, a lot of the phenomena occurring in real-time systems are relevant for the study of time-critical systems. As an example of this, in process control systems often continuous physical entities are involved such as temperature and volume. When such a system contains, for example, an analog circuit for monitoring the temperature, this has a time-continuous nature together with a continuous range of values (e.g. between -20 and +40 Centigrade). In modeling such systems, the usual discrete view of time as taken for digital systems is therefore not appropriate any more. Hence, apart from viewing time as discrete one should also allow a view of time as continuous (or at least dense) as in Newtonian physics. This has also its repercussion on the description of the execution of such a system (or rather how it develops) and how it can be observed. For discrete systems, execution consists of a number of observable state changes or transitions leading to a state-transition sequence. In the case of time-continuous systems, however, variables can change arbitrarily fast (think e.g. of pressure) and sequences cannot be used any more. A particular execution can only be described by recording at *each* moment the state of the system (so, such a generalized execution model considers functions from time to states). If one would maintain that observations can be made only at discrete moments, each observation contains only partial information. Only the whole set of possible observations of a particular execution can restore all information on that execution.

Summarizing, for time-critical systems quantitative temporal requirements play a dominant role. Furthermore, a discrete view of time and familiar execution models such as interleaving are not sufficient any more to handle all cases. Consequently, time-continuous models, respectively real parallelism or scheduling information should be incorporated.

6.3 How to Specify Time-Critical Systems

Like we did for message passing systems in section 3 of Chapter 5 let us specialize the requirements for a specification language in Chapter 2 to the case of time-critical systems.

Syntactical abstractness requires that the specification of temporal properties is stated only in terms of the events involved and the relevant quantity of time units.

The introduction of formal methods for time-critical systems has lagged behind that for other application areas. Most specification methods do not include constructs to express timing in a quantitative way and the few syntactical formalisms that include timing, lack formal semantics. It is thus not surprising that only a minority of suitable formal methods for time-critical systems have been developed. In fact, most of them were developed during the last few years. Some of these methods do not tackle all problems of time-critical systems but concentrate, for example, on discrete event systems. Several reasons can be given for the fact that formal methods for time-critical systems lag behind that for other application areas:

- because the timing requirements are much stricter for time-critical systems than for other systems, they impose more demands on the implementation technology; therefore, implementation concerns (e.g. processor speed) were dominant in the era before the explosive growth of computing power for microprocessors that started about ten years ago,

- the intrinsic complexity of typical time-critical systems makes it much more difficult to develop adequate formal methods,

- most researchers in theoretical computer science have considered real-time either as a special (though admittedly harder) case of concurrent systems, or as a topic whose study should be postponed until we understand basic concurrency better.

Layered development is not as dominant for time-critical systems as it is for message passing systems but still top-down and bottom-up techniques are important for specifying these systems in order to manage their inherent complexity.

6.4   Metric Temporal Logic

In this section we look at ways of reasoning with temporal logic about quantitative timing properties such as those mentioned in section 1. The standard models for temporal logic based on point structures involve a pure qualitative view of time by considering only a set of moments T together with the precedence relation $<$ (see Definition 3.2.1 in section 2 of Chapter 3). The question now is: what should be added to such point structures $(T, <)$ to be able to handle also quantitative temporal properties? Because the evaluation of formulas is dependent on a particular point in time (representing the present), we suggest that apart from the precedence relation between the present and other points in time also the distance between points in time is needed. Therefore we add a distance function d with the idea that $d(t,t')$ gives a measure as to how far t and t' are apart in time. The next question is: what conditions should be put on $<$ and d ? Because quantitative temporal properties relating different components of a system must necessarily refer to a global conception of time, we assume that the set of time points can be ordered in a global way. So, we suppose that the precedence relation $<$ is total (i.e. transitive, irreflexive and comparable). For the distance function d we suppose the usual topological conditions apart from the replacement of the triangular inequality by a conditional equality:

(d1) $d(t,t') = 0 \Leftrightarrow t = t'$

(d2) $d(t,t') = d(t',t)$

(d3) if $t < t' < t''$ then
$$d(t,t'') = d(t,t') + d(t',t'') \text{ and } d(t'',t) = d(t'',t') + d(t',t).$$

Next we should determine the range of d. There is no reason to choose the standard reals (in fact, Example 6.4.2 below shows the usefulness of non-archimedean ranges for d). As is apparent from the conditions (d1)–(d3) above we need a structure with addition and zero element. So, we suppose as range for d a structure $(\Delta, +, 0)$ where addition $+$ and constant 0 are restricted by:

(Δ1) $\delta + \delta' = \delta' + \delta$ (commutativity)

(Δ2) $(\delta + \delta') + \delta'' = \delta + (\delta' + \delta'')$ (associativity)

(Δ3) $\delta + 0 = \delta = 0 + \delta$ (unit 0)

(Δ4) $\delta + \delta' = \delta + \delta'' \Rightarrow \delta' = \delta''$

       and   (+ injective in both arguments)

         $\delta + \delta'' = \delta' + \delta'' \Rightarrow \delta = \delta'$

(Δ5) $\delta + \delta' = 0 \Rightarrow \delta = 0$ and $\delta' = 0$ (no negative elements)

(Δ6) $\exists \delta'' [\delta = \delta' + \delta'' \text{ or } \delta' = \delta + \delta'']$ (existence of absolute difference).

In these conditions the free variables should be universally quantified (we left this out for the sake of concise presentation). One can easily check the independence of these restrictions on $(\Delta, +, 0)$, that is, that none of these restrictions follows from the others, by means of appropriate examples in which five of these restrictions hold and the sixth fails. An example is $\Delta = \mathbb{IN} \cup \{e\}$ where we take over the standard addition for natural numbers supplemented by the following rules for the extra element e (which resembles 1):

$$e + e = 2, \quad e + 0 = 0 + e = e \quad \text{and} \quad e + n = n + e = n + 1 \quad \text{for } n \in \mathbb{IN} \setminus \{0\}.$$

This structure $(\Delta, +, 0)$ obeys all restrictions $(\Delta 1)$–$(\Delta 6)$ above except $(\Delta 4)$:

$$e + e = e + 1, \text{ but } e \neq 1.$$

In spite of their independence these restrictions nevertheless contain some redundancy (e.g. the second equality in $\Delta 3$ is added although this already follows from $\Delta 1$) in order to state the intended restriction fully also in the case when some of the other restrictions have been dropped. These conditions are motivated as follows. $(\Delta 1)$ is enforced by (d2) and (d3). One also needs to order Δ to compare different distances (think e.g. of the expression of maximal distance, see point 1 in section 1). To this end, first define

$$\delta \preceq \delta' := \exists \delta'' \, [\delta' = \delta + \delta''].$$

Such a δ'' is unique because of $(\Delta 4)$. Furthermore, $\Delta 2$ (transitivity) and $\Delta 3$ (reflexivity) make $\preceq$ a preorder. The corresponding irreflexive relation defined by

$$\delta \prec \delta' := \exists \delta'' \, [\delta'' \neq 0 \text{ and } \delta' = \delta + \delta'']$$

is in fact a total order (comparable by $\Delta 6$) with 0 as its least element (by $\Delta 5$).

This leads to the following notion.

Definition 6.4.1
A *metric point structure* is a two-sorted structure $(T, \Delta, <, d, +, 0)$ with signature $< \subseteq T \times T, \ d : T \times T \to \Delta, \ + : \Delta \times \Delta \to \Delta, \ 0 \in \Delta$ such that

(i) $<$ is total

(ii) d is surjective and satisfies (d1)–(d3)

(iii) $(\Delta, +, 0)$ satisfies $(\Delta 1)$–$(\Delta 6)$.

Δ and d are called the *metric domain* and the *temporal distance function*, respectively.

In (ii) surjectivity of d is demanded to get a nice correspondence between T and Δ. All these conditions on $<$ and d were motivated either by practical reasons (having a certain application area in mind) or by our wish to obtain a nice mathematical theory. Nevertheless, in some cases these conditions could be relaxed, for example it may be beneficial to allow a cluster of points having distance 0 to each other (deleting the only if case of condition d1). For the time being, we consider the above conditions as the most natural ones.

Example 6.4.2
Consider the following metric point structure.
$$T := \text{IN} \times \text{IN}$$
$$\Delta := \{0\} \times \text{IN} \cup \text{IN}^+ \times \mathbb{Z}$$
where IN, IN^+ and $\mathbb{Z}$ represent the natural numbers, the positive natural numbers, respectively the integers.
Define furthermore
$$(n, n') < (m, m') := n < m \text{ or } (n = m \text{ and } n' < m')$$

$$d((n, n'), (m, m')) := \begin{cases} (0, |\, n' - m' \,|) & \text{if } n = m \\ (m - n, m' - n') & \text{if } n < m \\ (n - m, n' - m') & \text{if } n > m \end{cases}$$

$$(n, z) + (n', z') := (n + n', z + z')$$
$$0 := (0, 0).$$

The following picture represents T together with its ordering $<$ (to be read from left to right):

$$\quad (0,0) \ (0,1) \ (0,2) \quad (1,0) \ (1,1) \ (1,2) \quad (2,0) \ (2,1) \ (2,2)$$

The idea is that the first component of T represents a kind of macro-time while the second component represents micro-time. It is easy to check that this example satisfies all conditions for a metric point structure and that the given Δ is non-archimedean. Such a time domain might seem unrealistic but closely resembles the synchrony hypothesis which is at the heart of the execution models for synchronous languages such as ESTEREL (see [BC 85]) and Statecharts (see [Har 87]): the macro-time represents external time while micro-time takes care of the ordering of events within a macro-time step (a kind of internal time). We refer to [Hui 91] for a treatment of such issues in the semantics of reactive systems.

Having determined what the new temporal models should be, we now must find appropriate temporal operators for reasoning about metric point structures. Before making such a choice we show how the modal operators **L** and **M** (see Chapters 3 and 4) can be transformed into metric operators:

$$\mathbf{L}_\delta^R \, \varphi(w) \ := \ \forall w' \in W[(wRw' \ \text{and} \ d(w,w') = \delta) \ \Rightarrow \ \varphi(w')]$$

$$\mathbf{M}_\delta^R \, \varphi(w) \ := \ \exists w' \in W[wRw' \ \text{and} \ d(w,w') = \delta \ \text{and} \ \varphi(w')].$$

Again $\mathbf{M}_\delta^R \equiv \overline{\mathbf{L}_\delta^R}$.

Now two obvious metric operators are $\mathbf{F}_\delta := \mathbf{M}_\delta^<$ and $\mathbf{P}_\delta := \mathbf{M}_\delta^>$ with their duals $\mathbf{G}_\delta \equiv \mathbf{L}_\delta^<$ and $\mathbf{H}_\delta \equiv \mathbf{L}_\delta^>$, respectively. For metric point structures other metric operators can be expressed with these two, for example, $\mathbf{D}_\delta \equiv \mathbf{P}_\delta \vee \mathbf{F}_\delta$, but to be able to express the requirements on the distance function d in an independent fashion (later in this section) we also introduce

$$\mathbf{D}_\delta \ := \ \mathbf{M}_\delta^{\neq}$$
$$\text{and}$$
$$\mathbf{E}_\delta \ := \ \mathbf{M}_\delta^{T \times T}.$$

Formally, we use the standard first-order language (including identity $=$) over $(\Delta, +, 0)$ whose terms t are used to form the metric operators $\mathbf{F}_t, \mathbf{P}_t, \mathbf{D}_t$ and $\mathbf{E}_t$. In the qualitative case (see section 2 of Chapter 4) the relation between $\mathbf{E}$ and $\mathbf{D}$ was given by $\mathbf{E}\varphi \equiv \varphi \vee \mathbf{D}\varphi$. In the quantitative case, the general semantic relation between $\mathbf{E}_\delta$ and $\mathbf{D}_\delta$ (following immediately from the definitions above) looks similarly:

$$\mathbf{E}_\delta \, \varphi(t) \ \Leftrightarrow \ (d(t,t) = \delta \ \text{and} \ \varphi(t)) \ \text{or} \ \mathbf{D}_\delta \, \varphi(t).$$

If we assume $d(t,t) = 0$ (which follows from d1) this reduces to the syntactic equivalence

$$\mathbf{E}_\delta \, \varphi \ \equiv \ \mathbf{D}_\delta \, \varphi \vee (\delta = 0 \wedge \varphi).$$

Now, in case $\delta \neq 0$ we find that $\mathbf{E}_\delta \equiv \mathbf{D}_\delta$. In case $\delta = 0$ and assuming $d(t,t') = 0 \ \Rightarrow \ t = t'$ (the other part of d1) we find that $\mathbf{D}_0 \, \varphi \equiv \ \bot$ and

$\mathbf{E}_0\,\varphi \equiv \varphi$. So, only in the case $\delta = 0$ the old equivalence $\mathbf{E}\varphi \equiv \varphi \vee \mathbf{D}\varphi$ is maintained.

$\mathbf{E}_\delta$ and $\mathbf{D}_\delta$ are not the only metric operators that are strongly related. A further pair is formed by $\mathbf{F}_\delta$ and $\mathbf{G}_\delta$ (and similarly for $\mathbf{P}_\delta$ and $\mathbf{H}_\delta$). If we assume (d1)–(d3) and comparability of $<$ (both are true for metric point structures) it is easy to see that $\mathbf{F}_\delta$ can indicate at most one point (i.e. $\forall t \neg \exists t' \exists t'' [t < t'$ and $t < t''$ and $t' \neq t''$ and $d(t,t') = d(t,t'') = \delta]$). Because $\mathbf{G}_\delta$ is the dual of $\mathbf{F}_\delta$ it must indicate the same point (if it exists). In fact, the existence of this point is exactly the difference between $\mathbf{F}_\delta$ and $\mathbf{G}_\delta$ ($\mathbf{F}_\delta$ asserts its existence while $\mathbf{G}_\delta$ does not) as is expressed by the syntactical equivalence

$$\mathbf{F}_\delta\,\varphi \equiv \mathbf{F}_\delta \top \wedge \mathbf{G}_\delta\,\varphi.$$

From the operators $\mathbf{F}_\delta$ and $\mathbf{P}_\delta$ several more metric operators can be derived:

$$\mathbf{F}_{<\delta}\,\varphi := \exists \delta'[0 \prec \delta' \prec \delta \wedge \mathbf{F}_{\delta'}\,\varphi]$$

$$\mathbf{P}_{<\delta}\,\varphi := \exists \delta'[0 \prec \delta' \prec \delta \wedge \mathbf{P}_{\delta'}\,\varphi]$$

$$\varphi\,\mathbf{until}_\delta\,\psi := \mathbf{F}_\delta\,\psi \wedge \mathbf{G}_{<\delta}\,\varphi$$

$$\varphi\,\mathbf{since}_\delta\,\psi := \mathbf{P}_\delta\,\psi \wedge \mathbf{H}_{<\delta}\,\varphi$$

where $\mathbf{G}_{<\delta}$ and $\mathbf{H}_{<\delta}$ are the duals of $\mathbf{F}_{<\delta}$ and $\mathbf{P}_{<\delta}$, respectively:

$$\mathbf{G}_{<\delta}\,\varphi := \neg\,\mathbf{F}_{<\delta}\,\neg\,\varphi$$

$$\mathbf{H}_{<\delta}\,\varphi := \neg\,\mathbf{P}_{<\delta}\,\neg\,\varphi.$$

Using these metric operators the five quantitative temporal properties of section 1 can be expressed in the following way:

1. maximal distance: $\mathbf{A}(p \rightarrow \mathbf{F}_{<\delta}\,q)$

2. exact distance: $\mathbf{A}(p \rightarrow \mathbf{F}_\delta\,q)$

3. minimal distance: $\mathbf{A}(p \rightarrow \neg\mathbf{F}_{<\delta}\,p)$

4. periodicity (with period δ): $\mathbf{E}p \wedge \mathbf{A}(p \rightarrow (\neg p\,\mathbf{until}_\delta\,p))$

5. bounded response time: $\exists\delta\,\mathbf{A}(p \rightarrow \mathbf{F}_{<\delta}\,q)$.

The one but last of these five properties gives periodicity towards the future
(the $\mathbf{E}p$ is needed to start the sequence off). Periodicity both towards past
and future can be expressed by

$$\mathbf{E}p \;\wedge\; \mathbf{A}(p \;\rightarrow\; ((\neg p \,\mathbf{until}_\delta\, p) \;\wedge\; (\neg p \,\mathbf{since}_\delta\, p))).$$

Note that the definition of $\mathbf{F}_{<\delta}$ uses quantification over Δ but this was already
essentially needed for the expression of bounded response time (see 5). As
stated above, besides constants from Δ (the δ in 1,2,3 and 4) we incorporate
the full first-order language over $(\Delta,+,0)$. Later on we will also consider
a fragment of metric temporal logic in which only constants from Δ are
allowed. The formula expressing maximal distance is strictly stronger than
the formula for bounded response time which on its turn is strictly stronger
than the formula $\mathbf{A}(p \rightarrow \mathbf{F}q)$ expressing temporal implication in qualitative
temporal logic. The latter fact can be illustrated by the following example.
Take $T := \mathsf{IN}$ with the standard ordering $<$ on IN and the standard distance
function $(\Delta = \mathsf{IN},\ d(m,n) = |m - n|)$. Let p be true precisely in elements
of $\{n(n+1)/2 - 1 \mid n > 0\}$ and q precisely in elements of $\{n(n+1)/2 + n - 1 \mid n > 0\}$. This choice of valuation ensures that the distance between the
n-th occurrence of p and the n-th occurrence of q will be n. Therefore, in
this metric model bounded response time will fail but qualitative temporal
implication still holds.

The ability to quantify over Δ gives metric temporal logic considerable
expressive power. For example, from the metric version of an operator the
corresponding qualitative operator can easily be derived by 'quantifying δ
away' as we will show below. Furthermore, for qualitative temporal logic
the operators **until** and **since** add expressive power (see [Kam 68]) but as
just shown their metric versions (and hence by quantifying δ away also their
qualitative versions) are expressible in metric temporal logic. These reduc-
tions of the operators **until** and **since** in metric temporal logic deal with
the equivalence of formulas over *models*. Like we did in Chapters 3 and 4
for classical modal and temporal logic, respectively polymodal logics with
inequality, we can ask which first-order conditions are definable by a formula
from metric temporal logic over *frames*. It turns out that *all* first-order conditions over *linear* orders are definable in metric temporal logic, as shown
below in Theorem 6.4.3. Because quantification over Δ contributes signifi-
cantly to the expressive power of metric temporal logic, we now study the
interplay between metric operators and quantification over Δ. We start with
the simple case of two existential quantifications (for the moment we return

to the more general case of metric modal operators and subsequently use the results for metric temporal logic):

$$\exists \delta \, \mathbf{M}_\delta \, \varphi(w) \quad \equiv \quad \exists w' \in W \, \exists \delta [wRw' \text{ and } d(w, w') = \delta \text{ and } \varphi(w')]$$
$$\equiv \quad \exists w' \in W [wRw' \text{ and } \varphi(w')] \equiv \mathbf{M} \, \varphi(w),$$

so $\exists \delta \, \mathbf{M}_\delta \equiv \mathbf{M}$. By duality also $\forall \delta \, \mathbf{L}_\delta \equiv \mathbf{L}$. The presence of two identical (either existential or universal) quantifiers is in itself not a sufficient explanation for these equivalences. For example, for classical temporal logic (see section 2 of Chapter 3) $\mathbf{HG}\varphi \equiv \mathbf{GH}\varphi$ is not valid because of the shifting of the reference point (consider e.g. IN in the point 0). In the present case, however, identical quantifications over the metric domain and over the set of worlds do not influence each other and hence can be interchanged.

More interesting are the cases of alternating quantifiers:

$$\forall \delta \, \mathbf{M}_\delta \, \varphi(w) \quad \equiv \quad \forall \delta \, \exists w' \in W [wRw' \text{ and } d(w, w') = \delta \text{ and } \varphi(w')].$$

For metric point structures $<$ is comparable and (d1) and (d3) hold. As we have seen above this implies that $\mathbf{F}_\delta$ and $\mathbf{G}_\delta$ are related by

$$\mathbf{F}_\delta \, \varphi \quad \equiv \quad \mathbf{F}_\delta \top \wedge \mathbf{G}_\delta \, \varphi.$$

So, when universally quantifying over δ (excluding $\delta = 0$ because $\mathbf{F}_0 \, \varphi \equiv \bot$ for all φ) we get

$$\forall \delta [0 \prec \delta \; \rightarrow \; \mathbf{F}_\delta \, \varphi] \quad \equiv \quad \forall \delta [0 \prec \delta \; \rightarrow \; \mathbf{F}_\delta \top] \wedge \mathbf{G} \, \varphi$$

(since $\forall \delta [0 \prec \delta \; \rightarrow \; \mathbf{G}_\delta \, \varphi] \equiv \mathbf{G} \, \varphi$). Dually we have

$$\exists \delta [0 \prec \delta \wedge \mathbf{G}_\delta \, \varphi] \quad \equiv \quad \exists \delta [0 \prec \delta \wedge \mathbf{G}_\delta \bot] \vee \mathbf{F} \, \varphi.$$

Note that $\forall \delta [0 \prec \delta \; \Rightarrow \; \mathbf{F}_\delta \top (t)]$ expresses the requirement that there exists for each $\delta \neq 0$ a point in the future with distance δ from t which is like surjectivity of d but now demanded locally (for t).

Besides quantification over metric operators we can look at special values of δ in $\mathbf{M}_\delta$ and $\mathbf{L}_\delta$ such as 0:

$$\mathbf{M}_0 \, \varphi(w) \quad \equiv \quad \exists w' \in W [wRw' \text{ and } d(w, w') = 0 \text{ and } \varphi(w')].$$

Assuming (d1) and taking $\varphi \equiv \top$ we get $\mathbf{M}_0 \, \varphi(w) \equiv wRw$. So,

$\mathbf{E} \, \mathbf{M}_0 \top$ expresses $\exists w \, wRw$ (existence of a reflexive world)

$\mathbf{A} \, \mathbf{M}_0 \top$ expresses $\forall w \, wRw$ (reflexivity)

and dually

$$\mathbf{E}\,\mathbf{L}_0 \perp \text{ expresses } \exists w\, \neg w R w \qquad \text{(existence of an irreflexive world)}$$
$$\mathbf{A}\,\mathbf{L}_0 \perp \text{ expresses } \forall w\, \neg w R w \qquad \text{(irreflexivity)}.$$

This example shows that a qualitative property (the existence of a reflexive world) is definable when using metric modal operators while it is not in its qualitative version $\text{PML}(R, \neq)$, see Proposition 4.2.4 in section 2 of Chapter 4. Turning again to metric temporal logic we can in fact prove:

Theorem 6.4.3 All first-order conditions over linear orders are definable in metric temporal logic.

Proof: The main problem in translating first-order conditions on $<$ into equivalent temporal formulas is caused by the possibility to compare in the first-order condition a 'new' variable (corresponding to a more recent reference point in time) with much 'older' variables such as the comparisons between z and x and between u and y in the example

$$\forall x\, \exists y > x\, \exists z < x\, \forall u\, (z < u < y \;\rightarrow\; u = x).$$

Qualitative temporal logics only allow a comparison between a new reference point in time and the most recent reference point before that. Our remedy against this difficulty works as follows. Let $\alpha(x_1, \ldots, x_n)$ be the first-order sentence (containing the bound variables $x_1, \ldots, x_n$) to be defined by a formula from MTL. First rewrite α in such a way that it only contains the atomic formulas $x_i < x_j$ and $x_i = x_j$ (for $1 \leq i, j \leq n$) and operators $\neg, \wedge$ and $\exists$. Furthermore, take care that each atomic formula in the scope of $\exists x_i$ indeed contains x_i (otherwise get the atomic formula outside this scope). The translation of the resulting first-order sentence into a formula from metric temporal logic is based on the following idea. For metric point structures, the comparison of different reference points in time can be accomplished by using the distance function as follows. All points in time are compared with a fixed reference point in time which is characterized by the fact that this point is the only point at which a certain proposition p holds (see below how this can be achieved). The first-order variables $x_1, \ldots, x_n$ are translated into variables $\delta_1, \ldots, \delta_n$ which represent the distance to the fixed reference point (where p holds) taking into account comparisons with other variables using $<$ and $>$ by the appropriate future and past metric operators. The translation of α into a formula from metric temporal logic is given by

$$\mathbf{A}((p \wedge \neg \mathbf{D}\, p) \;\rightarrow\; \mu_{(0,\ldots,0)}(\alpha))$$

where the procedure $\mu_{(s_1,\ldots,s_n)}$ is defined below. The prefix $\mathbf{A}((p \wedge \neg \mathbf{D}p) \rightarrow$ fixes the only point where p holds, the fixed reference point. To indicate the comparisons with the fixed reference point the procedure μ uses additional variables $s_1, \ldots s_n \in \{-, 0, +\}$ ($-$ indicates the past, 0 the present and $+$ the future). Initially $s_1, \ldots, s_n$ are all 0. μ is defined recursively as follows:

$$\mu_{\bar{s}}(\neg \alpha) \quad := \quad \neg \mu_{\bar{s}}(\alpha)$$
$$\mu_{\bar{s}}(\alpha \wedge \beta) \quad := \quad \mu_{\bar{s}}(\alpha) \wedge \mu_{\bar{s}}(\beta)$$
$$\mu_{\bar{s}}(\exists x_i\, \alpha) \quad := \quad \exists \delta_i\; \mathbf{E}[(\delta_i = 0 \wedge p \wedge \mu_{\bar{s}[0/s_i]}(\alpha))$$
$$\vee\ (0 \prec \delta_i \wedge \mathbf{F}_{\delta_i} p \wedge \mu_{\bar{s}[-/s_i]}(\alpha))$$
$$\vee\ (0 \prec \delta_i \wedge \mathbf{P}_{\delta_i} p \wedge \mu_{\bar{s}[+/s_i]}(\alpha))]$$

$$\mu_{\bar{s}}(x_i < x_i) \quad := \quad \bot$$
$$\mu_{\bar{s}}(x_i = x_i) \quad := \quad \top$$

$$\mu_{\bar{s}}(x_i < x_j) \quad := \quad \begin{cases} \mathbf{F}p & \text{if } s_j = 0 \\ \mathbf{FF}_{\delta_j} p & \text{if } s_j = - \\ \mathbf{FP}_{\delta_j} p & \text{if } s_j = + \end{cases}$$

$$\mu_{\bar{s}}(x_j < x_i) \quad := \quad \begin{cases} \mathbf{P}p & \text{if } s_j = 0 \\ \mathbf{PF}_{\delta_j} p & \text{if } s_j = - \\ \mathbf{PP}_{\delta_j} p & \text{if } s_j = + \end{cases}$$

$$\mu_{\bar{s}}(x_i = x_j) \quad := \quad \begin{cases} p & \text{if } s_j = 0 \\ \mathbf{F}_{\delta_j} p & \text{if } s_j = - \\ \mathbf{P}_{\delta_j} p & \text{if } s_j = + \end{cases}$$

where x_i in the last five cases (from $x_i < x_i$ onwards) is the bound variable belonging to the smallest enclosing existential quantification. ■

Finally, we look at axiomatizations for metric temporal logic. Completeness may be unattainable because of the very powerful quantification over Δ. By assuming an oracle for Δ *relative* completeness results might be obtained, however. Although completeness results are not so readily obtainable, all conditions in the definition of a metric point structure can easily be expressed:

(i) totality of $<$ is already expressible in $PML(<, >, \neq)$, see section 2 of Chapter 4

(ii) d surjective: $\forall \delta\ \mathbf{E}\, \mathbf{E}_\delta\, \top$
    (d1): $p \leftrightarrow \mathbf{E}_0\, p$
    (d2): $\forall \delta\ [(p \wedge \mathbf{E}_\delta\, q) \rightarrow \mathbf{E}_\delta\, (q \wedge \mathbf{E}_\delta\, p)]$
    (d3): $\forall \delta\ \forall \delta'\ [(\mathbf{F}_\delta\, \mathbf{F}_{\delta'}\, p \rightarrow \mathbf{E}_{\delta+\delta'}\, p) \wedge (\mathbf{P}_\delta\, \mathbf{P}_{\delta'}\, p \rightarrow \mathbf{E}_{\delta+\delta'}\, p)]$

(iii) $(\Delta 1)$–$(\Delta 6)$ can be directly formulated in terms of $+$, 0 and quantification over Δ.

To give an example of the four equivalences in (ii) we prove the first one.

> First suppose d is surjective. This means that for all $\delta \in \Delta$ there exist $t, t' \in T$ such that $d(t, t') = \delta$. Hence t verifies $\mathbf{E}_\delta \top$. Thus, $\forall \delta\, \mathbf{E}\, \mathbf{E}_\delta \top$ is true.
>
> Conversely, suppose $\forall \delta\, \mathbf{E}\, \mathbf{E}_\delta \top$ is true. Then for all $\delta \in \Delta$ there exists a $t \in T$ such that $\mathbf{E}_\delta \top$ is true in t, implying the existence of a $t' \in T$ at distance δ from t. Thus, $d(t, t') = \delta$, so d is surjective.

Instead of attempting to axiomatize metric temporal logic completely we can at least provide a sound axiomatization. A first proposal is:

Definition 6.4.4 The metric temporal logic proof system M consists of

0. the definitions

$$
\begin{aligned}
\exists \delta\, \varphi(\delta) &:= \neg\, \forall \delta\, \neg \varphi(\delta), \\
\mathbf{G}_t\, \varphi &:= \neg\, \mathbf{F}_t\, \neg\varphi, \\
\mathbf{H}_t\, \varphi &:= \neg\, \mathbf{P}_t\, \neg\varphi, \\
\mathbf{F}\, \varphi &:= \exists \delta\, \mathbf{F}_\delta\, \varphi, \\
\mathbf{G}\, \varphi &:= \forall \delta\, \mathbf{G}_\delta\, \varphi, \\
\mathbf{P}\, \varphi &:= \exists \delta\, \mathbf{P}_\delta\, \varphi, \\
\mathbf{H}\, \varphi &:= \forall \delta\, \mathbf{H}_\delta\, \varphi,
\end{aligned}
$$

1. a complete axiomatization of predicate logic including MP (Modus Ponens) and the following two rules ($\forall$-elimination, respectively $\forall$-introduction):

   a. to infer $\varphi(t)$ from $\forall \delta\, \varphi(\delta)$, where $\varphi(t)$ is the result of substituting the term t from the first-order structure $(\Delta, +, 0)$ properly (i.e. avoiding that any free variable of t becomes bound) for all occurrences of δ in $\varphi(\delta)$,

   b. to infer $\varphi \to \forall \delta\, \psi(\delta)$ from $\varphi \to \psi(t)$, where t is a term from the first-order structure $(\Delta, +, 0)$ that does not appear in $\varphi \to \forall \delta \psi(\delta)$,

2. the distribution axiom schemas and temporalization rules of the minimal temporal logic proof system K_t (see Definition 3.2.32 in section 2 of Chapter 3) for $\mathbf{G}_t$ and $\mathbf{H}_t$:

a. $\mathbf{G}_t\,(\varphi \rightarrow \psi) \rightarrow (\mathbf{G}_t\,\varphi \rightarrow \mathbf{G}_t\,\psi)$ and
$\mathbf{H}_t\,(\varphi \rightarrow \psi) \rightarrow (\mathbf{H}_t\,\varphi \rightarrow \mathbf{H}_t\,\psi)$,

b. to infer $\mathbf{G}_t\,\varphi$ from φ, and to infer $\mathbf{H}_t\,\varphi$ from φ,

3. the characterizations (i)–(iii) of the properties of a metric point structure above,

4. the already mentioned additional relationships between metric operators:

$$\mathbf{F}_t\,\varphi \leftrightarrow \mathbf{F}_t\,\top \wedge \mathbf{G}_t\,\varphi,$$
$$\mathbf{P}_t\,\varphi \leftrightarrow \mathbf{P}_t\,\top \wedge \mathbf{H}_t\,\varphi,$$

5. axiom schemas relating to arithmetic over the metric domain:

a. $\mathbf{F}_0\,\varphi \leftrightarrow \bot \leftrightarrow \mathbf{P}_0\,\varphi$

b. $\mathbf{F}_{t_1}\,\dot{\mathbf{F}}_{t_2}\,\varphi \leftrightarrow \mathbf{F}_{t_1}\,\top \wedge \mathbf{F}_{t_1+t_2}\,\varphi,$
$\mathbf{P}_{t_1}\,\dot{\mathbf{P}}_{t_2}\,\varphi \leftrightarrow \mathbf{P}_{t_1}\,\top \wedge \mathbf{P}_{t_1+t_2}\,\varphi,$
where metric operators are made reflexive in a similar way as for polymodal operators (see section 2 of Chapter 4), for example

$$\dot{\mathbf{F}}_t\,\varphi := (t = 0 \wedge \varphi) \vee \mathbf{F}_t\,\varphi.$$

c. $\mathbf{F}_{t_1}\,\mathbf{P}_{t_1+t_2}\,\varphi \leftrightarrow \mathbf{F}_{t_1}\,\top \wedge \dot{\mathbf{P}}_{t_2}\,\varphi,$
$\mathbf{P}_{t_1}\,\mathbf{F}_{t_1+t_2}\,\varphi \leftrightarrow \mathbf{P}_{t_1}\,\top \wedge \dot{\mathbf{F}}_{t_2}\,\varphi,$
$\mathbf{F}_{t_1+t_2}\,\dot{\mathbf{P}}_{t_1}\,\varphi \leftrightarrow \mathbf{F}_{t_1+t_2}\,\top \wedge \dot{\mathbf{F}}_{t_2}\,\varphi,$
$\mathbf{P}_{t_1+t_2}\,\dot{\mathbf{F}}_{t_1}\,\varphi \leftrightarrow \mathbf{P}_{t_1+t_2}\,\top \wedge \dot{\mathbf{P}}_{t_2}\,\varphi.$

From this proof system several interesting properties can be derived such as

(1) $\forall \delta\,\mathbf{G}\,\varphi(\delta) \leftrightarrow \mathbf{G}\,\forall \delta\,\varphi(\delta)$ and $\mathbf{F}\,\forall \delta\,\varphi(\delta) \rightarrow \forall \delta\,\mathbf{F}\,\varphi(\delta)$ (these follow from predicate logic and the definitions $\mathbf{G}\varphi \equiv \forall \delta'\,\mathbf{G}_{\delta'}\,\varphi$ and $\mathbf{F}\varphi \equiv \exists \delta'\,\mathbf{F}_{\delta'}\,\varphi$),

(2) $\forall \delta\,\mathbf{F}_\delta\,\varphi \leftrightarrow \forall \delta\,\mathbf{F}_\delta\,\top \wedge \mathbf{G}\,\varphi$ by predicate logic and clause 4 in the definition of M above,

(3) $\mathbf{F}_t\,\mathbf{P}_t\,\varphi \leftrightarrow \mathbf{F}_t\,\top \wedge \varphi$ by taking $t_2 = 0$ in the first axiom schema of clause 5c in the definition of M above

and similarly for the mirror images (obtained by exchanging $\mathbf{G}$ with $\mathbf{H}$ and $\mathbf{F}$ with $\mathbf{P}$).

The next properties are important enough to derive them as theorems of M. In these derivations MP abbreviates Modus Ponens and M followed by a number indicates the corresponding clause in the definition of M above.

Proposition 6.4.5 $\vdash_M$ $\mathbf{G}_t(\varphi \wedge \psi) \leftrightarrow \mathbf{G}_t \varphi \wedge \mathbf{G}_t \psi$

Proof: This theorem of M can be derived as follows.

1. $\varphi \wedge \psi \rightarrow \varphi$ (propositional logic)

2. $\mathbf{G}_t(\varphi \wedge \psi \rightarrow \varphi)$ (1,M2b)

3. $\mathbf{G}_t(\varphi \wedge \psi) \rightarrow \mathbf{G}_t \varphi$ (2,MP,M2a)

4. $\mathbf{G}_t(\varphi \wedge \psi) \rightarrow \mathbf{G}_t \psi$ (analogous to 1–3)

5. $\mathbf{G}_t(\varphi \wedge \psi) \rightarrow \mathbf{G}_t \varphi \wedge \mathbf{G}_t \psi$ (3,4)

6. $\varphi \rightarrow (\psi \rightarrow \varphi \wedge \psi)$ (propositional logic)

7. $\mathbf{G}_t(\varphi \rightarrow (\psi \rightarrow \varphi \wedge \psi))$ (6,M2b)

8. $\mathbf{G}_t \varphi \rightarrow \mathbf{G}_t (\psi \rightarrow \varphi \wedge \psi)$ (7,MP,M2a)

9. $\mathbf{G}_t \varphi \rightarrow (\mathbf{G}_t \psi \rightarrow \mathbf{G}_t(\varphi \wedge \psi))$ (8,MP,M2a)

10. $\mathbf{G}_t \varphi \wedge \mathbf{G}_t \psi \rightarrow \mathbf{G}_t(\varphi \wedge \psi)$ (9,propositional logic)

11. $\mathbf{G}_t(\varphi \wedge \psi) \leftrightarrow \mathbf{G}_t\varphi \wedge \mathbf{G}_t\psi$ (5,10)

■

This was not very surprising since this holds also for the non-metric case: $\mathbf{G}(\varphi \wedge \psi) \leftrightarrow \mathbf{G}\varphi \wedge \mathbf{G}\psi$ (and indeed the derivation above uses only clause 2 of M which stems from the minimal temporal logic proof system K_t). However, in contrast with the non-metric case we have also the following:

Proposition 6.4.6 $\vdash_M$ $\mathbf{F}_t(\varphi \wedge \psi) \leftrightarrow \mathbf{F}_t \varphi \wedge \mathbf{F}_t \psi$

Proof: This theorem of M can be derived as follows.

1. $\mathbf{F}_t(\varphi \wedge \psi) \leftrightarrow \mathbf{F}_t \top \wedge \mathbf{G}_t(\varphi \wedge \psi)$ (M4)

2. $\mathbf{G}_t(\varphi \wedge \psi) \leftrightarrow \mathbf{G}_t \varphi \wedge \mathbf{G}_t \psi$ (Proposition 6.4.5)

3. $\mathbf{F}_t(\varphi \wedge \psi) \leftrightarrow \mathbf{F}_t \top \wedge \mathbf{G}_t \varphi \wedge \mathbf{G}_t \psi$ (1,2)

4. $\mathbf{F}_t \top \wedge \mathbf{G}_t \varphi \leftrightarrow \mathbf{F}_t \varphi$ (M4)

5. $\mathbf{F}_t \top \wedge \mathbf{G}_t \psi \leftrightarrow \mathbf{F}_t \psi$ (M4)

6. $\mathbf{F}_t(\varphi \wedge \psi) \leftrightarrow \mathbf{F}_t\,\varphi \wedge \mathbf{F}_t\,\psi$ (3,4,5)

∎

The only part of K_t that we did not take over concerns the tense mixing axiom schemas $\varphi \rightarrow \mathbf{GP}\varphi$ and $\varphi \rightarrow \mathbf{HF}\varphi$. These are however theorems of M, for instance the first one:

Proposition 6.4.7 $\vdash_M \varphi \rightarrow \mathbf{GP}\varphi$

Proof: This theorem of M can be derived as follows.

1. $\mathbf{F}_t\,\top \vee \neg\,\mathbf{F}_t\,\top$ (propositional logic)

2. $\mathbf{F}_t\,\top \wedge \varphi \rightarrow \mathbf{F}_t\,\mathbf{P}_t\,\varphi$ (M5c)

3. $\mathbf{F}_t\,\mathbf{P}_t\,\varphi \leftrightarrow \mathbf{F}_t\,\top \wedge \mathbf{G}_t\,\mathbf{P}_t\,\varphi$ (M4)

4. $\mathbf{F}_t\,\top \wedge \varphi \rightarrow \mathbf{G}_t\,\mathbf{P}_t\,\varphi$ (2,3)

5. $\neg\,\mathbf{F}_t\,\neg\,\mathbf{P}_t\,\varphi \leftrightarrow \neg(\mathbf{F}_t\,\top \wedge \mathbf{G}_t\,\neg\,\mathbf{P}_t\,\varphi)$ (M4)

6. $\neg\,\mathbf{F}_t\,\top \rightarrow \neg\,\mathbf{F}_t\,\neg\,\mathbf{P}_t\,\varphi$ (5)

7. $\neg\,\mathbf{F}_t\,\top \rightarrow \mathbf{G}_t\,\mathbf{P}_t\,\varphi$ (6,M0)

8. $\varphi \rightarrow \mathbf{G}_t\,\mathbf{P}_t\,\varphi$ (1,4,7)

9. $\varphi \rightarrow \forall\delta\,\mathbf{G}_\delta\,\mathbf{P}_\delta\,\varphi$ (8,M1b)

10. $\varphi \rightarrow \forall\delta\,\mathbf{G}_\delta\,\exists\delta'\,\mathbf{P}_{\delta'}\,\varphi$ (9, $\delta' = \delta$)

11. $\varphi \rightarrow \mathbf{GP}\varphi$ (10,M0)

∎

Another possibility is to eliminate the quantification over Δ by only allowing constants from Δ. Such a fragment of metric temporal logic could be based on the following eight temporal operators: $\mathbf{until}_{<\delta}$, $\mathbf{until}_\delta$, $\mathbf{until}_{>\delta}$, $\mathbf{until}$, $\mathbf{since}_{<\delta}$, $\mathbf{since}_\delta$, $\mathbf{since}_{>\delta}$, $\mathbf{since}$ where δ may be any constant from Δ. Notice that we now included the qualitative operators $\mathbf{until}$ and $\mathbf{since}$ because these can no longer be obtained by quantification over their metric equivalents. An alternative way to look at these qualitative operators is to see them as special metric operators $\mathbf{until}_{<\infty}$ and $\mathbf{since}_{<\infty}$ as is done in

[HW 89]. In this view ∞ is not an element of Δ but it is added to $\prec$ as its greatest element. [HW 89] is also interesting for another reason: it shows how MTL can be embedded as an assertion language into a compositional proof system.

Another look at the constants δ from Δ is to consider them as programs from a kind of dynamic logic (see [Har 84]) by defining

$$[\delta] := \{(t,t') \mid d(t,t') = \delta\}$$

with the following additional program structure

$$0: \quad \text{the 'skip' program}$$
$$+: \quad \text{sequential composition ;}$$

and the property that all programs are *deterministic*:

$$\mathbf{F}_\delta\,\varphi \wedge \mathbf{F}_\delta\,\psi \;\rightarrow\; \mathbf{F}_\delta\,(\varphi \wedge \psi)$$

(cf. Proposition 6.4.6 above). This connection with dynamic logic deserves further investigation.

In the same way as indicated in section 3 of Chapter 3 for $L(\mathbf{until}, \mathbf{since})$ we can introduce global variables and quantification over them in order to reason about (possibly infinite) data domains like that of messages. This will be illustrated in the next section.

6.5   Specification Examples

In this section we illustrate the application of metric temporal logic to time-critical systems by a series of examples. The first three examples treat some simple, but characteristic, pure real-time phenomena: pure time-out, a watchdog timer monitoring a processor and the wait/delay statement. The next four examples combine features of message passing and time-critical systems. Example four concerns a terminal adaptor where the speed of the incoming data is higher than the speed of the outgoing data. In example five a synchronous and an asynchronous input are mixed into one synchronous output. Example six treats an abstract transmission medium. Real-time communication constructs like send and receive with time-out are the subject of example seven. Example eight deals with process control systems in which continuously changing entities play an important role. Finally, example nine implements the reliable transmission media used in the third layer of the communication network of Example 3 of section 6 of Chapter 5 by using a time-out and retransmission facility over unreliable communication links.

We use the same priority of operators as in section 6 of Chapter 5. Also (as we did in section 6 of Chapter 5) we assume in our specifications linearity of the ordering and succession towards future. This involves the qualitative part of metric temporal logic. For the quantitative part we assume local surjectivity of the temporal distance function d, that is, we assume $\forall \delta \, [0 \prec \delta \rightarrow \mathbf{F}_\delta \top]$. An important consequence of this is $\mathbf{F}_\delta \equiv \mathbf{G}_\delta$ for all $\delta \neq 0$ since $\mathbf{F}_\delta \, \varphi \equiv \mathbf{F}_\delta \top \wedge \mathbf{G}_\delta \, \varphi$ (see section 4). The standard metric point structures that we have in mind use respectively the natural numbers, the integers, the (non-negative) rational numbers and the (non-negative) real numbers for the time domain T and the non-negative part of T for Δ where $<, +$ and 0 have the standard interpretation for these number systems and d is the absolute difference. For example, one of the standard metric point structures is

$$(\mathbb{Z}, \mathsf{IN}, <, d, +, 0)$$

where $<$ is the standard ordering on $\mathbb{Z}$, $+$ the standard addition on IN, 0 the standard constant from IN and d is defined by

$$d(z, z') \; := \; |z - z'|.$$

For the specification examples in this section we need two additional qualitative temporal operators above those introduced in section 6 of Chapter 5. First we need a reflexive version of **since** which we will denote by **<u>since</u>**. Semantically it corresponds to replacing every $<$ in the definition of **since** by $\leq$. Syntactically this can be achieved by the definition

$$\varphi_1 \, \underline{\mathbf{since}} \, \varphi_2 \; := \; (\varphi_1 \wedge \varphi_2) \vee (\varphi_1 \wedge \varphi_1 \mathbf{since}(\varphi_1 \wedge \varphi_2)).$$

Apart from this binary reflexive operator we also need a unary operator denoted by $\mathbf{J}$ representing that its argument has *just* become true:

$$\mathbf{J}\varphi \; := \; \varphi \wedge (\mathbf{P}\varphi \rightarrow \neg\varphi \, \mathbf{since} \, \neg\varphi).$$

(We thank Job Zwiers for the discussion leading to this more concise representation of this operator than the one contained in [Koy 89].) This definition can be explained as follows. Apart from the obvious first part demanding that φ holds at the current moment, this formula describes that there was a period immediately before (how small it may be) such that φ was false in

that period. Note that for a formula φ that is true on the rationals and false
on the irrational numbers $\mathbf{J}\varphi$ is never true (this corresponds to our intuition
that φ changes its truth value infinitely fast and hence cannot have become
just true).

In our examples we will encounter periodicity requirements. Uncondi-
tional periodicity of an event e with period δ can be formulated by

$$periodic(e, \delta) \; := \; e \; \rightarrow \; \neg\, e \,\mathbf{until}_\delta\, e.$$

Furthermore, conditional periodicity can be defined by adding a condition c
to the antecedent:

$$periodic(e, \delta, c) \; := \; e \wedge c \; \rightarrow \; \neg\, e \,\mathbf{until}_\delta\, e.$$

In applying metric temporal logic to practical examples the metric domain Δ should be associated with a time unit relevant for that application,
usually second or a derivative thereof. However, in principle other time units
such as number of shaft rotations are allowed too. Connected with this is
the translation of elements of data domains that represent time units into
elements of Δ. We will represent this translation by a function δ. For example, when the data domain represents milliseconds and Δ counts in seconds
than we can take $\delta(t) = \frac{t}{1000}$. In case the data domain has more structure,
one may want to impose additional conditions on δ, for example, when the
data domain is ordered monotonicity of δ with respect to this ordering and
when the data domain incorporates addition distributivity of δ with respect
to this addition. The most simple case occurs when the data domain can be
embedded in the metric domain. In such a case it suffices to take for δ simply
the embedding mapping.

In examples three and seven we look at statements from concurrent programming languages such as CHILL ([CHILL 80]) or Ada ([Ada 83]). For
expressing the semantics of programming languages we use location variables
l and location predicates at and $after$. The first assumption on locations is
that being simultaneously at and after the same location is impossible (being simultaneously at different locations in different processes or tasks is of
course possible):

$$\mathbf{L1} \qquad \neg\,(at(l) \wedge after(l)).$$

Locations are special data elements and as such we can impose on them
the Unique Identification assumption. However, being present at a certain

location is not instantaneous, but has some duration (an extended event), so
the uniqueness of locations is expressed by

L2 $at(l) \rightarrow at(l)$ **unless** $(after(l) \wedge \mathbf{G} \neg at(l))$.

As we did in section 6 of Chapter 5 we leave out universal quantifications
over the data domains in the specifications.

We take the same attitude as in section 6 of Chapter 5 with regard to the
specification of the interface.

6.5.1   Example 1: Pure Time-out

One of the most common and easiest real-time constructs is the time-out. A
time-out is generated at the end of a period (whose length is determined by
the value by which the timer was set) in which a certain event (think of the
signal resetting the timer) has not occurred. Time-outs are widely used in
real-time systems to safeguard one part of a system against malfunction of
another part. Let the event be e and the time-out value δ, then the time-out
on e after δ can be defined by the MTL formula

$$time\_out(e, \delta) := \neg \mathbf{P}_{<\delta}\, e.$$

So, a time-out on e after δ is generated if and only if e has not occurred
during the last δ time units. Notice that in this representation the setting of
the timer is considered irrelevant. If we want to incorporate this, however,
let set and $reset$ be the event setting, respectively resetting, the timer, then
a time-out with period δ can be described by

$$\neg\, reset\ \mathbf{since}_{\delta}\ set.$$

6.5.2   Example 2: Watchdog Timer

This example concerns a pure real-time system, a watchdog timer. A processor is monitored by a timer, the watchdog. The processor sets the timer by
a signal $enable(t)$ and it should reset the timer by a $reset$ signal each time
before the timer expires (cf. the previous example). When the processor does
not succeed in resetting the timer in time, the processor will be stopped by a
$halt$ signal from the watchdog. At any time, the processor and the watchdog
timer can be restarted by an $initiate$ signal from the environment (e.g. an
operator pushing a button). After an $initiate$ signal a new period of enabling

and resetting the timer starts. Once the timer is set with $enable(t)$ after an initiate signal, the time-out period cannot be changed (and thus every subsequent $enable(t')$ signal is ignored) until the next $initiate$ signal. Figure 6.1 summarizes this state of affairs.

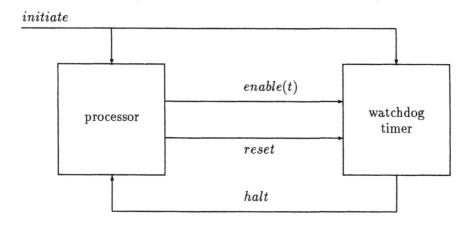

Figure 6.1: Watchdog Timer

We assume that the enable-line obeys the no simultaneous input assumption (otherwise the time-out period could be unknown):

$$enable(t) \land enable(t') \rightarrow t' = t.$$

To identify the first $enable(t)$ after an initiate signal we define

$$firstenable(t) := enable(t) \land (\neg \exists t' \, enable(t')) \text{ since } initiate.$$

The only essential thing to be specified is the generation of the $halt$ signal. This is characterized by a period bounded by $firstenable(t)$ (timer set) and a $halt$ signal (timer stopped) in which:

1. no $initiate$ and no $halt$ signal occurred during this whole period (no $halt$ signal since we want at most one $halt$ signal to occur between two $initiate$ signals),

2. no $reset$ occurred during the last t time units of this period.

The generation of a $halt$ signal can then be specified by a nested since formula:

$$halt \leftrightarrow \exists\, t\, [\, t > 0 \wedge (\neg\, initiate \wedge \neg\, halt \wedge \neg\, reset)\, \mathbf{since}_{\delta(t)}$$
$$((\neg\, initiate \wedge \neg\, halt)\, \underline{\mathbf{since}}\, firstenable(t))]$$

where δ transfers an element from the data domain of *enable* to an element of the metric domain Δ (see the introduction of this section).

6.5.3 Example 3: Wait/delay Statement

This example treats the wait statement or delay statement as occurring in concurrent programming languages such as CHILL ([CHILL 80]) or Ada ([Ada 83]). See the introduction of this section for the way we use locations to express the semantics of programming languages. By $wait(l)$ we denote that l is the location of a wait statement and $waitvalue(l)$ denotes the specified waitvalue of that wait statement. The semantics of a wait statement is then specified by

$$\mathbf{J}\, at(l) \wedge wait(l) \rightarrow at(l)\, \mathbf{until}_{\delta(waitvalue(l))}\, after(l).$$

Remark 6.5.1 For the **J**-operator and the function δ transferring elements from a data domain to elements of the metric domain Δ, see the introduction of this section.

Remark 6.5.2 The *just*-operator **J** is essential because we need to characterize the moment when $at(l)$ has just become true. This moment can be seen as the time when a timer is set with a time-out of $\delta(waitvalue(l))$ time units. Leaving out **J** would then correspond to a continuous resetting of this timer which contradicts the demand that $after(l)$ will be true after $\delta(waitvalue(l))$ time units.

Remark 6.5.3 Being present at a location takes some time so the wait statement cannot be passed in 0 time units. In other words, even if the waitvalue is 0 the function δ will take care that this is mapped to a positive number to account for the time it takes to transfer control (cf. Appendix A in [KSRGA 85] concerning this problem for the Ada delay statement).

Remark 6.5.4 If also an infinite waitvalue is allowed we add the following axiom for this special case:

$$at(l) \wedge wait(l) \wedge waitvalue(l) = \infty \rightarrow \mathbf{G}\, at(l).$$

6.5.4   Example 4: Terminal Adaptor

This example is a mixture of message passing and real-time. It concerns a simplified terminal adaptor. On one side bytes are received from a data link operating on 512 bytes/second. On the other side bytes are transmitted to a terminal with a rate of 300 bytes/second. The adaptor has a buffering capacity of N_1 bytes and it prevents buffer overflow through sending *stop* and *start* signals to the data link as soon as the buffer becomes more than 80% full, respectively more than 80% empty. It is assumed that after the sending of a *stop* signal at most N_2 bytes are sent by the data link (of course N_2 should be small compared to N_1). The data link may resume sending bytes only after it has received a *start* signal. Let $in(b)$ denote the reception of byte b from the high-speed data link and $out(b)$ the transmission of byte b to the terminal. The above is summarized in Figure 6.2.

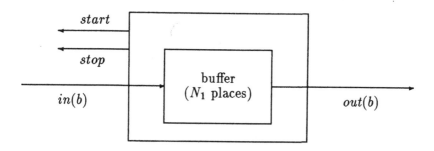

Figure 6.2: Terminal Adaptor

The terminal adaptor is a perfect FIFO message passing system, so we suppose:

Unique Identification (MP1) for *in*,

No Creation and finite speed (MP2′) for *out* with respect to *in*,

perfectness (MP3′),

no simultaneous input and output (MP4a,b) for *in* and *out*,

FIFO ordering (MP5).

Additionally, the terminal adaptor obeys some real-time restrictions. First define

$$buffered(b) := \dot{\mathbf{P}}\,in(b) \wedge \neg \dot{\mathbf{P}}\,out(b)$$

to express that byte b is at the moment contained in the buffer of the terminal adaptor. We assume that transmission of bytes to the terminal is irregular (i.e. aperiodic), but within $\frac{1}{300}$ of a second:

$$buffered(b) \;\rightarrow\; \mathbf{F}_{<\frac{1}{300}} \,\exists\, b' \; out(b').$$

Because the buffer respects FIFO ordering this can be strengthened to

$$buffered(b) \;\wedge\; \neg\,\exists\, b'[buffered(b') \;\wedge\; \dot{\mathbf{P}}(in(b) \;\wedge\; \mathbf{P}\; in(b'))] \;\rightarrow$$
$$\neg\,\exists\, b'\; out(b') \; \mathbf{until}_{<\frac{1}{300}} \; out(b)$$

where $\varphi \; \mathbf{until}_{<\delta} \; \psi$ is of course defined by

$$\exists\delta' \,[0 \prec \delta' \prec \delta \;\wedge\; \varphi \; \mathbf{until}_{\delta'} \; \psi].$$

The strengthened axiom above can be derived as an instance (by taking $\varphi \equiv \exists b'\, out(b')$ and $\psi \equiv out(b)$) from

$$\neg\,\varphi \; \mathbf{until} \; \psi \;\wedge\; \mathbf{F}_{<\delta}\,\varphi \;\rightarrow\; \neg\,\varphi \; \mathbf{until}_{<\delta} \; \psi$$

(where $\neg\varphi \; \mathbf{until} \; \psi$ stems from the part about the FIFO ordering).

We now proceed with the other side, the reception of bytes from the data link. Define

$$stopped \;:=\; (\neg\, start) \; \mathbf{since} \; stop$$

$$start\_stop\_interference \;:=\; \Diamond_{<\frac{1}{512}} (stop \vee start)$$

(where $\Diamond_{<\delta}\,\varphi$ is defined by $\varphi \vee \mathbf{F}_{<\delta}\,\varphi$) to indicate that the reception was stopped (a $stop$ signal was issued and since then no $start$ signal has been issued), respectively a period (of length $\frac{1}{512}$) in which reception is interfered by issuing a $stop$ or $start$ signal. We can now specify the regular reception of bytes from the data link with period $\frac{1}{512}$, unless reception was stopped or interfered by a $stop$ or $start$ signal:

$$in(b) \;\wedge\; \neg\, stopped \;\wedge\; \neg\, start\_stop\_interference \;\rightarrow$$
$$\neg\,\exists\, b'\; in(b') \; \mathbf{until}_{\frac{1}{512}} \;\exists\, b'\; in(b').$$

Remark 6.5.5 This axiom represents a conditional periodicity requirement. Therefore, the above axiom can also be written as

$$periodic(\exists\, b'\; in(b'), \frac{1}{512}, \neg\, stopped \;\wedge\; \neg\, start\_stop\_interference).$$

(Recall from predicate logic that $\forall x[(P(x) \wedge Q) \rightarrow R]$ is equivalent with $(Q \wedge \exists x\, P(x)) \rightarrow R$ when Q and R do not contain x free.)

Remark 6.5.6 Note that $\neg\,(\neg\, start \textbf{ since } stop) \wedge \neg\,\Diamond_{<\frac{1}{512}}\,(stop \vee start)$ is equivalent with $\neg\,\mathbf{F}_{\frac{1}{512}}\,(\neg\, start \textbf{ since } stop) \wedge \neg\,\Diamond_{<\frac{1}{512}}\, start$ (the latter formulation was used in [KKZ 87]).

After a *stop* signal the data link need not immediately stop sending bytes (it can still send at most N_2 bytes). Nevertheless, the reception of bytes remains regular in such a period. To enforce this we also demand backward periodicity after the first byte after the last *start* signal:

$$in(b) \rightarrow \neg\,\exists\, b'\, in(b') \textbf{ since } start \vee \neg\,\exists\, b'\, in(b') \textbf{ since}_{\frac{1}{512}} \exists\, b'\, in(b').$$

After a *stop* signal at most N_2 bytes can be sent by the data link:

$$\neg\, start \textbf{ since}_{>\frac{N_2}{512}} stop \rightarrow \neg\,\exists\, b\, in(b)$$

where $\varphi \textbf{ since}_{>\delta} \psi$ is defined by

$$\exists\delta'\,[\delta \prec \delta' \wedge \varphi \textbf{ since}_{\delta'} \psi].$$

At last we should specify the generation of the *start* and *stop* signals. For convenience we assume that N_1 is divisible by 5. To indicate the situation that the buffer is for at least 80% full, respectively at least 80% empty, we define

$$almostfull := \exists b_1 \cdots \exists b_{\frac{4}{5}N_1+1}\, [\bigwedge_{\substack{i,j=1 \\ i<j}}^{\frac{4}{5}N_1+1} b_i \neq b_j \wedge \bigwedge_{i=1}^{\frac{4}{5}N_1+1} buffered(b_i)]$$

$$almostempty := \neg\,\exists b_1 \cdots \exists b_{\frac{1}{5}N_1}\, [\bigwedge_{\substack{i,j=1 \\ i<j}}^{\frac{1}{5}N_1} b_i \neq b_j \wedge \bigwedge_{i=1}^{\frac{1}{5}N_1} buffered(b_i)].$$

Remark 6.5.7 N_1 is a fixed (constant) parameter in this specification so that the sequence of existential quantifiers in front of these formulas can be replaced by a sequence of fixed length.

Remark 6.5.8 When one allows the use of auxiliary data structures such as a queue, one simply could refer to the length of the queue representing the buffer. However, we consider the use of auxiliary data structures against the requirement of syntactical abstractness for specification languages (see Chapter 2 and section 5 of Chapter 5). When one decides to use only logical and temporal operators combined with quantification over and equality in the data domain (in this case bytes), a bit more complex definitions like the ones above are unavoidable.

Now we should specify that the *start* and *stop* signals will be generated *as soon as* the buffer becomes (again) almost full, respectively almost empty. To express the as soon as aspect, we use the just-operator **J** (see the introduction of this section):

$$start \ \leftrightarrow \ \mathbf{J} \ almostempty$$

$$stop \ \leftrightarrow \ \mathbf{J} \ almostfull.$$

As one can see from these two axioms the *start* and *stop* signals are not essential and, using these two axioms, can be consequently replaced in the previous axioms by their equivalent right-hand sides. In other words, this specification can be given in a more abstract way only in terms of *in* and *out* without the implementation-oriented signals *start* and *stop*! This phenomenon occurs because we see systems as black boxes and hence only specify the outside (see Chapter 2), but on the other hand overviewing this outside from all sides (seeing the *whole* environment). In case of the terminal adaptor, the *start* and *stop* signals are essential from an implementation viewpoint because the data link cannot see from its position how the other side (the terminal) is doing, in particular how fast the terminal adaptor transmits bytes at that side. Because the data link does not have this information, it is not able to stop in right time and start sending bytes again when necessary *by itself*.

6.5.5   Example 5: Mixing Synchronous and Asynchronous Input

In this example we specify an object with two inputs and one output. The original informal specification is contained in [DHJR 85]:

> The object has two inputs and one output. The output and one of the inputs respectively send and receive data in packets at regular intervals. The remaining input is asynchronous, i.e. data appears at undetermined times.

The data packets which arrive at the synchronous input may be full or empty, and the object may only output data by forwarding packets from the synchronous input or filling an empty packet with data from the asynchronous input. All packets have the same size.

This is represented in Figure 6.3.

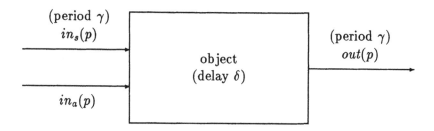

Figure 6.3: Mixing Synchronous and Asynchronous Input

The object, like the terminal adaptor of Example 4, has a mixture of message passing and real-time features. It seems the intention of the informal specification above that the periods of the output and the synchronous input are the same (in the picture represented by $\gamma > 0$). If the period of the output would be shorter than that of the synchronous input, the output will have to create packets at a certain moment and this violates the No Creation assumption for message passing systems. If, on the other hand, the period of the output would be longer than that of the synchronous input, the output cannot keep pace and packets will be lost eventually. As we interpret the above informal specification this seems not intended because that specification suggests that the object functions as a *perfect* message passing system. Furthermore, we assume finite speed for the passing of packets. Because of the synchrony of the output and one of the inputs this leads to a fixed delay $\delta > 0$. This delay δ represents a kind of processing time to pass or possibly fill a packet. The message passing aspect of the object is somewhat unusual because only one output is coupled to two inputs. The most important input is, however, the synchronous one and the asynchronous one only functions in exceptional cases (an empty packet on the synchronous input). Therefore, the following message passing properties hold between the two inputs and the output: No Creation and finite speed hold between the output and both inputs, FIFO holds for the output and the synchronous input while perfectness only holds for non-empty packets on the synchronous input. These message passing properties will be a consequence of stronger real-time properties given below. We do assume no simultaneous input and output:

$$in_s(p) \land in_s(p') \rightarrow p' = p$$

$$in_a(p) \land in_a(p') \rightarrow p' = p$$

$$out(p) \land out(p') \rightarrow p' = p.$$

Also unique identification is supposed. Because the inputs are not separated like in Example 2 in section 6 of Chapter 5, but are mixed in this case, we must not only assume unicity for both inputs separately but also for the inputs between each other:

$$in_s(p) \lor in_a(p) \rightarrow \neg \mathbf{D} \left(in_s(p) \lor in_a(p) \right)$$

$$\neg \left(in_s(p) \land in_a(p) \right).$$

Recall from section 2 of Chapter 5 that the No Creation assumption on message passing systems consisted of two parts: no new messages and no duplicates. The no new messages part will follow from the real-time requirements below, but the no duplicates part is independent from the message passing relation between the output and the two inputs described above. So, we demand for the output:

$$out(p) \rightarrow \neg \mathbf{D} \, out(p).$$

We can now turn to the real-time requirements of the object. Using the abbreviation $periodic(e, \delta)$ defined in the introduction of this section, regularity of the output, respectively synchronous input, is required by

$$periodic(\exists \, p' \, in_s(p'), \gamma)$$
$$\text{and}$$
$$periodic(\exists \, p' \, out(p'), \gamma).$$

The following two real-time requirements concern perfectness with a delay of δ differentiating the cases of a non-empty and empty packet on the synchronous input:

$$in_s(p) \land \neg \, empty(p) \rightarrow \mathbf{F}_\delta \, out(p)$$

$$in_s(p) \land empty(p) \rightarrow \mathbf{F}_\delta \left(out(p) \lor \exists \, p' \, [out(p') \land \mathbf{P} \, in_a(p')] \right).$$

Remark 6.5.9 The latter axiom allows that a packet arrives on the asynchronous input at the very last moment. This is not quite in accordance with the idea that the delay δ represents a kind of processing time to pass or possibly fill a packet. More tailored towards this idea would be the axiom

$$in_s(p) \wedge empty(p) \rightarrow \mathbf{F}_\delta \, out(p) \vee \exists p' \, [\dot{\mathbf{P}} \, in_a(p') \wedge \mathbf{F}_\delta \, out(p')],$$

that is, getting the **P**-operator out of the scope of the $\mathbf{F}_\delta$.

Remark 6.5.10 Both axioms together (with the obvious change in case the alteration suggested in Remark 6.5.9 is taken into account) guarantee that

$$in_s(p) \rightarrow \mathbf{F}_\delta(out(p) \vee \exists p'[out(p') \wedge \mathbf{P}in_a(p')]),$$

so in particular

$$in_s(p) \rightarrow \mathbf{F}_\delta \, \exists p' \, out(p').$$

Remark 6.5.11 Because of Remark 6.5.10 and regularity of the output with period γ, the axiom for regularity of the synchronous input with period γ can be weakened to

$$in_s(p) \rightarrow \mathbf{F}_\gamma \, \exists p' \, in_s(p').$$

The reason is that $in_s(p) \wedge \mathbf{F}_{<\gamma} \, \exists p' \, in_s(p')$ implies by Remark 6.5.10

$$\mathbf{F}_\delta(\exists p' \, out(p') \wedge \mathbf{F}_{<\gamma} \, \exists p' \, out(p'))$$

which contradicts the regularity of the output with period γ.

Our last axiom ensures that output does not start too early, to be precise only after a delay δ after the first packet on the synchronous input:

$$\neg \, \exists p \, \dot{\mathbf{P}} \, in_s(p) \rightarrow \mathbf{F}_\delta \, \neg \, \exists p' \, \dot{\mathbf{P}} \, out(p').$$

An equivalent formulation of this axiom looks backwards:

$$\exists p' \, \dot{\mathbf{P}} \, out(p') \rightarrow \mathbf{P}_\delta \, \exists p \, \dot{\mathbf{P}} \, in_s(p).$$

Now we can show that the remaining message passing properties are implied by the above real-time requirements. First, an obvious strengthening of Remark 6.5.10 gives

$$in_s(p) \rightarrow \mathbf{F}_\delta((out(p) \wedge \mathbf{P}_\delta \, in_s(p)) \vee \exists p'[out(p') \wedge \mathbf{P} \, in_a(p')]).$$

So, each packet on the synchronous input leads after a delay δ to the output of either that packet or an earlier packet from the asynchronous input. Since the synchronous input and the output have the same period γ these packets caused by the synchronous input make up for all packets on the output from a delay δ after the first packet on the synchronous input (there can be no packets in between since the output is regular and there can be no packets simultaneously with those generated by the synchronous input because no simultaneous output is assumed). The last axiom ensures that before a delay δ after the first packet on the synchronous input there can be no packet on the output. Thus, the only packets on the output are those generated by a packet on the synchronous input as formulated by the above formula. Inspecting this formula we immediately can conclude no creation of new packets and finite speed since either $out(p) \wedge \mathbf{P}_\delta\, in_s(p)$ or $out(p') \wedge \mathbf{P}\, in_a(p')$ holds. In fact, we showed that $out(p) \rightarrow \mathbf{P}_\delta\, in_s(p) \vee \mathbf{P}\, in_a(p)$. No duplication of packets was already formulated separately and takes care that a packet from the asynchronous input cannot be taken twice to fill an empty packet from the synchronous input. FIFO ordering for packets from the synchronous input follows because the above implies that a packet from the synchronous input will be output after a delay δ or not at all as formulated by the formula

$$in_s(p) \;\rightarrow\; \mathbf{F}_\delta\, out(p) \;\vee\; \mathbf{A}\, \neg\, out(p).$$

Perfectness for non-empty packets of the synchronous input follows already solely from the axiom about non-empty packets at the synchronous input.

6.5.6 Example 6: Abstract Transmission Medium

A transmission medium can be considered as a message passing system where the input and output are called transmit, respectively arrive and the data consists of signals. We assume the following aspects of message passing systems: unique identification of signals, no creation of signals and finite transmission speed, basic liveness, no simultaneous input and output. As given in Example 1 of section 6 of Chapter 5 these can be formulated respectively by:

$$transmit(s) \;\rightarrow\; \neg\, \mathbf{D}\, transmit(s)$$

$$arrive(s) \;\rightarrow\; \mathbf{P}\, transmit(s) \wedge \neg\, \mathbf{D}\, arrive(s)$$

$$\mathbf{G}\,\mathbf{F}\,\exists\, s\; transmit(s) \;\rightarrow\; \mathbf{F}\,\exists\, s\; arrive(s)$$

$$transmit(s) \wedge transmit(s') \;\rightarrow\; s' = s$$

$$arrive(s) \land arrive(s') \rightarrow s' = s.$$

The characteristic feature of the transmission medium on top of being a particular kind of message passing system is the requirement that it is not too lazy, that is, there exists a fixed period γ in which the transmission medium attempts to transmit at least one signal (successfully or not). So, when there are no other signals to be transmitted, γ represents the maximum time for which the attempt to transmit a signal can be delayed. Such a requirement is needed to enable higher-level protocols to time-out on signals sent but not yet received and start retransmission. This is formulated by

$$\exists \gamma \; \mathbf{A} \; (\exists s \; [\dot{\mathbf{P}} \; transmit(s) \; \land \; \neg \dot{\mathbf{P}} \; arrive(s)] \; \rightarrow$$
$$\exists s' \; [\dot{\mathbf{P}} \; transmit(s') \; \land \; \neg \dot{\mathbf{P}} \; arrive(s') \; \land \; \mathbf{G}_{>\gamma} \neg arrive(s')])$$

where $\mathbf{G}_{>\delta} \, \varphi$ is defined by

$$\forall \delta' \; [\delta \prec \delta' \; \rightarrow \; \mathbf{G}_{\delta'} \, \varphi].$$

In the axiom above s' represents one signal which has been attempted to transmit in a particular period γ. If this transmission has been successful, $\mathbf{F}_{\leq\gamma} \, arrive(s')$ holds (where $\mathbf{F}_{<\delta} \, \varphi$ is defined by $\mathbf{F}_{\delta} \, \varphi \lor \mathbf{F}_{<\delta} \, \varphi$), otherwise $\mathbf{A} \neg \, arrive(s')$ holds. To prove this we note the following. Since $\mathbf{A} \neg \varphi$ is equivalent over linear orders with $\neg \dot{\mathbf{P}} \, \varphi \land \mathbf{G} \neg \varphi$ and $\neg \dot{\mathbf{P}} \, arrive(s')$ is given, it is sufficient to prove

$$\mathbf{G}_{>\gamma} \neg \, arrive(s') \; \rightarrow \; \mathbf{F}_{\leq\gamma} \, arrive(s') \lor \mathbf{G} \neg \, arrive(s').$$

Now, this is an instance of $\mathbf{G}_{>\gamma} \neg \psi \rightarrow \mathbf{F}_{\leq\gamma} \psi \lor \mathbf{G} \neg \psi$ which is a theorem of M as is shown by the following derivation:

1. $\exists \delta \; [0 \prec \delta \preceq \gamma \land \mathbf{F}_{\delta} \, \psi] \; \leftrightarrow \; \mathbf{F}_{\leq\gamma} \, \psi$ \hfill (definition $\mathbf{F}_{\leq\gamma} \, \psi$)

2. $\neg \forall \delta \; [0 \prec \delta \preceq \gamma \; \rightarrow \; \mathbf{G}_{\delta} \neg \psi] \; \leftrightarrow \; \mathbf{F}_{\leq\gamma} \, \psi$ \hfill (1,predicate logic,$M0$)

3. $\forall \delta \; [\gamma \prec \delta \rightarrow \mathbf{G}_{\delta} \neg \psi] \land \forall \delta \; [0 \prec \delta \preceq \gamma \rightarrow \mathbf{G}_{\delta} \neg \psi] \rightarrow \forall \delta \; [0 \prec \delta \rightarrow \mathbf{G}_{\delta} \neg \psi]$
   (predicate logic)

4. $\mathbf{G}_0 \neg \psi \; \leftrightarrow \; \top$ \hfill ($M5a$)

5. $\forall \delta \; [0 \prec \delta \; \rightarrow \; \mathbf{G}_{\delta} \neg \psi] \; \leftrightarrow \; \forall \delta \; \mathbf{G}_{\delta} \neg \psi$ \hfill (4,predicate logic)

6. $\forall \delta \; [0 \prec \delta \; \rightarrow \; \mathbf{G}_{\delta} \neg \psi] \; \leftrightarrow \; \mathbf{G} \neg \psi$ \hfill (5,$M0$)

7. $\forall \delta \; [\gamma \prec \delta \; \rightarrow \; \mathbf{G}_{\delta} \neg \psi] \land \forall \delta \; [0 \prec \delta \preceq \gamma \; \rightarrow \; \mathbf{G}_{\delta} \neg \psi] \; \rightarrow \; \mathbf{G} \neg \psi$ \hfill (3,6)

8. $\mathbf{G}_{>\gamma} \neg \psi \;\rightarrow\; \neg \forall \delta \, [0 \prec \delta \preceq \gamma \;\rightarrow\; \mathbf{G}_\delta \neg \psi] \;\vee\; \mathbf{G} \neg \psi$
   (7,definition $\mathbf{G}_{>\gamma} \neg \psi$)

9. $\mathbf{G}_{>\gamma} \neg \psi \;\rightarrow\; \mathbf{F}_{\leq \gamma} \psi \;\vee\; \mathbf{G} \neg \psi$ (2,8)

In this example we needed quantification over the metric domain.

6.5.7   Example 7: Real-Time Communication Constructs

In this example we describe asynchronous message passing by means of the
send and receive constructs. Our specific form of the send and receive constructs is inspired by CHILL (see [CHILL 80]). The send construct has an
associated signal which represents the data to be sent. Each signal has a
unique destination and every signal sent will eventually reach its destination.
The receive construct consists of a selection of signals that it may accept.
The selection is between signals that have been sent to the process to which
this receive construct belongs (that must be their destination), that have arrived and that have not been selected before. After a choice has been made,
control transfers to the corresponding part of the receive construct. So, for
a receive construct we can differentiate two phases:

1. wait (possibly forever) for a signal that can be accepted (one of the
   listed selection possibilities),

2. choose one of the acceptable signals and take the branch of that accepted signal.

In case of a timed receive construct the possibility of a time-out is added that
restricts the time the receiving process is going to wait for a signal matching
one of its selection possibilities to arrive. For real-time applications the communication constructs of (asynchronous) send and timed receive are the most
useful choices because they do not lead to deadlock possibilities (the sender
continues and the receiver times out). The send and receive constructs are
high-level communication primitives and are usually implemented on a network providing reliable communication by using time-out and retransmission
for unreliable transmission media like those of Example 6 (see also Example
9). Notice that send and receive resemble *in*, respectively *out*, of a perfect
message passing system. The main difference, however, is that the receiver

explicitly accepts signals at times *chosen by itself*. In other words: the possibility to output a message is under control of the environment instead of the system. We start by specifying the effect of a send statement:

$$at(l) \wedge send(l) \rightarrow at(l) \textbf{ until } after(l).$$

We use similar conventions about locations as we used in Example 3. $send(l)$ indicates that the location l contains a send statement. This axiom simply states that the send statement takes some finite time, and this is exactly the essence of an asynchronous send: the sender just continues in contrast with synchronous communication such as a rendezvous in Ada (see [Ada 83]). The signal that is the result of the send statement at location l will be represented by the function $signal(l)$. An alternative for this would be to put this explicitly in the predicate $send$, but in that case it should be additionally stated that only *one* signal is generated for each send statement:

$$send(l, s) \wedge send(l, s') \rightarrow s' = s.$$

We prefer the use of the predicate $send(l)$ and the function $signal(l)$ because then it is implicit that a send statement can generate only one signal. The fact that a signal s is sent can be expressed by

$$sent(s) \; := \; \exists\, l\, [\mathbf{J}\, after(l) \wedge send(l) \wedge signal(l) = s].$$

Here we use the just-operator to indicate that the moment of sending coincides precisely with the moment that the send statement has just been passed. Because send statements can be executed simultaneously at different places (locations in different processes) in the program, and similarly for receive statements, we cannot suppose the no simultaneous input assumption. We want the data passed to be unique, so we must demand that simultaneously executed send statements generate different signals:

$$send(l) \wedge send(l') \wedge signal(l) = signal(l') \rightarrow l' = l.$$

We now turn to the receiving side. As we indicated above, the message passing relation between the sender and the receiver is somewhat non-standard because the receiver chooses the time to make a selection between acceptable signals. This selection process is also a special one: only certain signals can be accepted. This is expressed by the predicate $selectable(s, l)$. There are several choices for the definition of this predicate depending on the intended possibilities to select signals, but the signal s should at least conform (either syntactically or semantically) to one of the possible choices of that particular receive statement (i.e. the one at location l) and the destination of s should be the process in which this receive statement (i.e. the location l) occurs. With a receive statement at location l and a signal s we associate the special location $choice(s, l)$ representing the location where control is transferred to when signal s is chosen to be accepted at l. For these special locations $choice(s, l)$ we again impose a uniqueness assumption:

$$choice(s, l) = choice(s', l') \rightarrow s' = s \wedge l' = l.$$

A signal s can be chosen to be accepted at l if it is selectable, has been sent and was not chosen before. So define

$$choosable(s, l) := selectable(s, l) \wedge \mathbf{P} \, sent(s) \wedge \neg \mathbf{P} \, \exists l' \, at(choice(s, l')).$$

The fact that $\mathbf{P} \, \exists l' \, at(choice(s, l'))$ models that a signal s has been chosen before depends crucially on the uniqueness assumption for the locations $choice(s, l)$. To see this, consider the following program with three processes:

$$\begin{aligned}
P_1 &:: \quad SEND\ 0\ TO\ P_3 \\
P_2 &:: \quad SEND\ 0\ TO\ P_3 \\
P_3 &:: \quad RECEIVE
\end{aligned}$$

$$\begin{aligned}
&0: \ldots \\
&1: \ldots\ ;
\end{aligned}$$

$$RECEIVE$$

$$\begin{aligned}
&0: \ldots \\
&1: \ldots
\end{aligned}$$

Let s_1 and s_2 be the signals sent from P_1 and P_2 respectively, and l one of the two receive statements in P_3, then

$$choice(s_i, l) = choice(s_{3-i}, l) \qquad \text{for } 1 \leq i \leq 2.$$

So, if s_1 is accepted in P_3 first, $\mathbf{P} \, \exists l' \, at(choice(s_2, l'))$ will hold although s_2 has not been chosen yet.

To arrive at a location $choice(s, l)$, s must have been choosable at l:

$$\mathbf{J} \, at(choice(s, l)) \rightarrow choosable(s, l).$$

The (non-timed) receive statement can now be described by the following two axioms:

$$at(l) \wedge receive(l) \rightarrow at(l) \ \mathbf{unless} \ \exists s \, at(choice(s, l))$$

$$at(l) \wedge receive(l) \wedge \exists s \, choosable(s, l) \; \rightarrow \; \mathbf{F} \, \exists \, s' \, at(choice(s', l)).$$

In the case of a timed receive statement there is the additional possibility to transfer control to the special else-location after timervalue (cf. the waitvalue of a wait statement in Example 3) time units have elapsed. Combined with the two axioms above for the non-timed case this leads to the axiom

$$\mathbf{J} \, at(l) \wedge \, timedreceive(l) \; \rightarrow$$
$$at(l) \; \mathbf{until}_{\delta(timervalue(l))} \; at(else(l)) \; \vee$$
$$at(l) \; \mathbf{until}_{<\delta(timervalue(l))}$$
$$((at(l) \wedge \exists \, s \, choosable(s, l)) \; \mathbf{until} \; \exists \, s' \, at(choice(s', l))).$$

Note that the choice to take the else-branch is always possible because it is not observable whether a signal has arrived at its destination or not. In other words, we know nothing about the speed of the reliable communication network. It would be realistic to impose an upper bound on the time for signals to arrive (the maximum transmission time). In that case the else-branch can only be taken if we add that there could not have arrived a signal within timervalue time units. This can be done by adding the following conjunct to the first clause of the disjunction in the axiom above (*maxtt* represents the maximum transmission time):

$$\wedge \, \neg \, \exists \, s[selectable(s, l) \; \wedge \; \neg \, \mathbf{P} \, \exists \, l' \, at(choice(s, l')) \; \wedge$$
$$\mathbf{F}_{\delta(timervalue(l))} \; \mathbf{P}_{>maxtt} \, sent(s)]$$

where $\mathbf{P}_{>\delta} \, \varphi$ is defined by

$$\exists \delta' \, [\delta \prec \delta' \; \wedge \; \mathbf{P}_{\delta'} \, \varphi].$$

In the same way one can introduce a minimum transmission time by incorporating such a *mintt* in the definition of *choosable(s, l)*:

$$selectable(s, l) \; \wedge \; \mathbf{P}_{>mintt} \, sent(s) \; \wedge \; \neg \, \mathbf{P} \, \exists \, l' \, at(choice(s, l')).$$

A (timed) receive statement can choose between several signals to accept. A fairness assumption can be added for these choices, relating to the locations *choice(s, l)*.

6.5.8 Example 8: Continuously Changing State Variables

In the previous examples we concentrated on events since these are very important for time-critical systems. In case state variables also play an important role, for instance in case of process control systems, it is still often the case that not the variable itself is the dominant feature but a certain event or condition involving this state variable. A typical example is a continuous physical variable like temperature. (Note: continuous here does not refer so much to the possibility of allowing a continuous range of values for temperature, but to the fact that temperature may change continuously in time.) Usually we are not interested in the absolute value of this state variable as such but more in the fact whether it stays within certain bounds, for example the system should only react when the condition $temperature \leq maxtemp$ becomes false. An instance where such a condition is crucial is a process where this condition leads to irreversible phenomena (such as a chemical chain reaction) that occur immediately. Whenever such a catastrophe occurs, emergency measures (such as a 'shutdown') are required. Suppose that $reactime$ is the required reaction time and that $shutdown$ is the required reaction, then such a requirement can be specified in metric temporal logic by

$$\mathbf{J}(temperature > maxtemp) \ \rightarrow \ \mathbf{F}_{<reactime} \ shutdown.$$

Here we use the just-operator to catch the exact moment when the condition $temperature \leq maxtemp$ changes from true to false.

One could argue that this behavior can be captured by a discrete system that takes samples of temperature and that uses only two events, $catastrophic$ and $critical$, that have the following correspondence:

$$catastrophic \ \equiv \ temperature \ > \ maxtemp$$

$$critical \ \equiv \ maxtemp \ - \ temperature \ < \ \epsilon$$

where ϵ is a constant depending on the rate of change of temperature and the chosen sample time. The point is that even if the possible rate of change is exactly known on beforehand and the chosen sample time is sufficiently fast, it cannot be determined in the case that temperature has a value inbetween $maxtemp - \epsilon$ and $maxtemp$ whether event $catastrophic$ will occur. For that reason, if one wants to stay at the safe side, this entails that a shutdown should be performed whenever the event $critical$ (instead of $catastrophic$) occurs. In other words, the threshold has been lowered from the occurrence of the event $catastrophic$, as originally intended, to the event $critical$. In temporal logic the above can be rephrased by the validity of

$$catastrophic \ \rightarrow \ \mathbf{P} \ critical$$

but not

$$critical \; \rightarrow \; \mathbf{F} \; catastrophic.$$

So, the conclusion must be that discrete systems can only model continuously changing state variables to a certain extent. Therefore, if one wants to specify the behavior of such state variables completely, a temporal logic that can reason about continous time domains is essential. Metric temporal logic caters for this possibility.

6.5.9   Example 9: Implementing Reliable Communication

This example continues Example 3 of section 6 of Chapter 5 by providing an implementation of the reliable transmission media used in layer 3 of that example (see e.g. Figure 5.6). This implementation is based on acknowledgements and uses retransmission when a time-out on the receipt of an acknowledgement occurs. In detail this works as follows.

Let $ack(p)$ denote the acknowledgement of p and let $\alpha > 0$ be the time-out period for acknowledgements. Because an acknowledgement of a packet travels from the receiver of the packet to the sender of that packet we deal here with a two-way transmission medium similar to the two-way channel of Example 2 of section 6 of Chapter 5 (see Figure 5.3). For convenience we denote one side of the medium by i and the other side by $3 - i$ (where $i \in \{1, 2\}$). Like we did in Example 6 we will call the input and output of packets transmit, respectively arrive. Since the transmission medium is unreliable it may lose packets. On the other hand it obeys the following aspects of message passing systems: no creation of packets and finite transmission speed, basic liveness, no simultaneous input and output, FIFO ordering. In terms of Example 1 of section 6 of Chapter 5 we assume the axioms **MP2a'**, **MP3**, **MP4a,b**, respectively **MP5**:

$$arrive_{3-i}(p) \; \rightarrow \; \mathbf{P} \; transmit_i(p)$$

$$\mathbf{G} \, \mathbf{F} \, transmit_i(p) \; \rightarrow \; \mathbf{F} \, arrive_{3-i}(p)$$

$$transmit_i(p) \wedge transmit_i(p') \; \rightarrow \; p' = p$$

$$arrive_{3-i}(p) \wedge arrive_{3-i}(p') \; \rightarrow \; p' = p$$

$$arrive_{3-i}(p) \wedge \mathbf{P} \, arrive_{3-i}(p') \; \rightarrow \; \mathbf{P} \, (transmit_i(p) \wedge \mathbf{P} \, transmit_i(p')).$$

Note that we cannot assume unique identification in the form of **MP1** because packets may get lost and have to be retransmitted. Furthermore, basic liveness now is strengthened to refer to a single packet because we must guarantee that each packet eventually arrives after a number of retransmissions. Also the axiom for FIFO ordering uses the strict versions of the **P**-operator because we have also assumed finite speed and no simultaneous input. The most important additional axiom is the following one:

$$transmit_i(p) \rightarrow$$
$$\mathbf{H} \neg transmit_i(p) \vee$$
$$(\neg arrive_i(ack(p)) \wedge \neg transmit_i(p)) \textbf{ since}_\alpha transmit_i(p).$$

This axiom states that whenever a packet is transmitted either it is the first time or it is retransmitted because an acknowledgement did not arrive in the time-out period α. This implies that a packet is retransmitted only if it is necessary (this replaces the unique identification assumption) as is illustrated by two consequences of the above axiom:

$$transmit_i(p) \rightarrow \mathbf{G}_{<\alpha} \neg transmit_i(p)$$

$$transmit_i(p) \wedge \mathbf{F}_{<\alpha} arrive_i(ack(p)) \rightarrow \mathbf{G} \neg transmit_i(p).$$

Next we have to specify the relation between the arrival of a packet and the sending of its acknowledgement. This is done in the following two axioms:

$$arrive_{3-i}(p) \rightarrow \mathbf{F} \, transmit_{3-i}(ack(p))$$

$$transmit_{3-i}(ack(p)) \rightarrow \mathbf{P} \, arrive_{3-i}(p).$$

Finally we have to specify the retransmission of a packet whenever no acknowledgement has been received in the time-out period α:

$$transmit_i(p) \wedge \neg \mathbf{F}_{<\alpha} arrive_i(ack(p)) \rightarrow \mathbf{F}_\alpha transmit_i(p).$$

The one and only crucial property to prove now, perfectness (i.e. that each packet when first transmitted will eventually arrive), is formulated by:

$$transmit_i(p) \wedge \mathbf{H} \neg transmit_i(p) \rightarrow \mathbf{F} \, arrive_{3-i}(p).$$

We prove this as follows.

> Suppose $transmit_i(p) \wedge \mathbf{H} \neg transmit_i(p)$. Then either $\exists n \, \mathbf{F}_{<n\alpha} arrive_i(ack(p))$ or $\mathbf{G} \neg arrive_i(ack(p))$. First we show that the second case leads to a contradiction. This second case implies $\forall n > 0 \, \mathbf{F}_{n\alpha} transmit_i(p)$ because of the retransmission

axiom. This implies $\mathbf{G\,F}\,transmit_i(p)$, so by basic liveness also $\mathbf{G\,F}\,arrive_{3-i}(p)$. The axioms relating the arrival of a packet and the sending of its acknowledgement now give $\mathbf{GF}transmit_{3-i}(ack(p))$, so again by basic liveness $\mathbf{G\,F}\,arrive_i(ack(p))$. However, this is a clear contradiction with the assumption of the second case. Therefore, we may assume the first case: $\exists n\,\mathbf{F}_{<n\alpha}\,arrive_i(ack(p))$. By the axiom formulating no creation of new packets this leads to $\exists\,n\,\mathbf{F}_{<n\alpha}\,\mathbf{P}\,transmit_{3-i}(ack(p))$. By the axioms relating the arrival of a packet and the sending of its acknowledgement this gives now $\exists\,n\,\mathbf{F}_{<n\alpha}\,\mathbf{P}\,\mathbf{P}\,arrive_{3-i}(p)$. Therefore we may conclude $\mathbf{E}\,arrive_{3-i}(p)$. Proving that $\dot{\mathbf{P}}\,arrive_{3-i}(p)$ leads to a contradiction gives us the desired conclusion $\mathbf{F}\,arrive_{3-i}(p)$. This contradiction can be derived easily: suppose $\dot{\mathbf{P}}\,arrive_{3-i}(p)$, then by the axiom formulating no creation of new packets we get $\dot{\mathbf{P}}\,\mathbf{P}\,transmit_i(p)$, but this immediately is in contradiction with the part $\mathbf{H}\,\neg\,transmit_i(p)$ of the perfectness property we have just proven now.

When relating the above axioms with Example 3 of section 6 of Chapter 5 one may easily get confused by the fact that the transmission primitives there are also called *transmit* and *arrive* although they have different properties. The easiest way of understanding the relationship between the *transmit* of that example and the *transmit* of this example is by making retransmission explicit. This can be done by an additional primitive *retransmit*. For example the last axiom above formulating the resending of packets whenever no acknowledgement has been received in time now splits into two axioms:

$$transmit_i(p) \,\wedge\, \neg\,\mathbf{F}_{<\alpha}\,arrive_i(ack(p)) \;\to\; \mathbf{F}_\alpha\,retransmit_i(p)$$

$$retransmit_i(p) \,\wedge\, \neg\,\mathbf{F}_{<\alpha}\,arrive_i(ack(p)) \;\to\; \mathbf{F}_\alpha\,retransmit_i(p).$$

Of course this can also be combined into one axiom:

$$(transmit_i(p) \,\vee\, retransmit_i(p)) \,\wedge\, \neg\,\mathbf{F}_{<\alpha}\,arrive_i(ack(p)) \;\to\;$$
$$\mathbf{F}_\alpha\,retransmit_i(p).$$

The relationship between the *arrive* of Example 3 of section 6 of Chapter 5 and the *arrive* of this example can be given in a similar way.

6.6   Conclusions

We end this chapter with some conclusions.

We extended temporal logic with metric operators derived from their qualitative polymodal versions described in Chapter 4. We showed how these metric operators could be usefully applied to the formal specification of time-critical systems. [Bur 84] section 6 contains an alternative proposal for metric temporal logic where time is structured as an ordered Abelian group. From a philosophical viewpoint, the idea that duration of time is expressed as an element of the time domain itself seems unnatural. Also technically, the natural addition on a time domain may not be sufficient for determining the distance between any two points, as is exemplified by the points $(0, 1)$ and $(1, 0)$ in Example 6.4.2 of section 4. When only interested in qualitative aspects of distances, however, Tarski's qualitative geometry ([Tar 69]) suggests models $(T, <, E)$ where E is the equidistance-predicate ($Exyuv$ if and only if x and y have the same distance as u and v). An interesting question connecting this approach with metric temporal logic is: how should $<$ and E be axiomatized to describe models $(T, <, E)$ that allow a *representation* in terms of our metric point structures such that $Exyuv \Leftrightarrow d(x, y) = d(u, v)$? Another alternative for expressing quantitative timing properties is dynamic logic (see [Har 84]) with one atomic program 'successor' S. But, already for the expression of bounded response time we need an *infinitary* dynamic logic ([Gol 82]):

$$\bigvee_n [S^*](p \rightarrow \bigvee_{i<n} < S^i > q).$$

This approach is only suitable for *discrete* structures, but our philosophy behind metric temporal logic required that the qualitative fragment concerning *all* point structures should be nicely embedded. This makes sense in practice too, because time-critical systems may contain non-discrete elements such as analog devices for handling continuous physical entities like temperature (see section 2 and Example 8 of section 5).

The semantics of metric temporal logic is based on points. As already indicated in section 4 of Chapter 3 an alternative approach is based on intervals instead. In the case of metric temporal logic this makes not much of a difference since intervals relating to quantitative timing elements are easily expressible in terms of the (point based) temporal operators. For instance, suppose we want to define an operator that expresses that a formula will be true somewhere in the interval between l and u time units from now, then this operator can be immediately defined in metric temporal logic by

$$\mathbf{F}_{(l,u)} \varphi := \exists \delta \, [l \prec \delta \prec u \wedge \mathbf{F}_\delta \varphi].$$

In fact, we can even do without quantification in this case:

$$\mathbf{F}_{(l,u)}\, \varphi \;:=\; l \prec u \,\wedge\, \mathbf{F}_l\, \mathbf{F}_{<\delta}\, \varphi$$

where δ corresponds with $u - l$, that is, that element of Δ such that $l + \delta = u$ which exists because $l \prec u$. Operators for closed, half-open and unbounded intervals can be defined in a similar manner.

The list of examples showed how several types of time-critical systems can be specified with metric temporal logic, ranging from very simple real-time constructs and systems to combined message passing/real-time systems and semantics for real-time communication constructs of concurrent programming languages. The resulting specifications are elegant and rather directly formalize our intuition about the timing aspects of for instance real-time systems.

Chapter 7

Summary and Concluding Remarks

In this monograph we develop a temporal logic for reasoning about message passing and time-critical systems and illustrate the resulting specification method by numerous examples. It is built on several papers that appeared between 1983 and 1990 ([KVR 83], [KR 85], [Koy 87], [KKZ 87], [KKZ 88], [KKZ 89], [Koy 90]). This research started at the author's practice period at Philips Telecommunication Industries from September 1982 till June 1983. The result was a paper ([KVR 83]) describing how the CHILL real-time asynchronous communication primitives *SEND* and *RECEIVE* could be described axiomatically in temporal logic. Being a first attempt, it contained several misconceptions. One of them was that time was considered as (a distinguished) part of the state and that a state change could occur without increasing the time component. More successful contributions of [KVR 83] were the use of past operators to obtain elegant specifications and the introduction of a powerful quantitative temporal operator (corresponding to **until**$_\delta$ of section 4 of Chapter 6). Examples 3 and 7 of section 5 of Chapter 6 show how such an axiomatization of the CHILL primitives would look like in the current formalism. One month after the presentation of that paper, in September 1983, a workshop was held in Cambridge where several specification formalisms (presented by Hoare, Lamport, Milner etcetera) were tested on the same set of ten examples (see the proceedings [DHJR 85]). In our contribution ([KR 85]) already some improvements were made: time was still part of the state, but now each state change necessarily increased time. The past operators and the quantitative **until**-operator again proved to be suitable, but on the other hand the next-operator (see section 4 of Chapter

3) was used only for the purpose of obtaining irreflexive operators. How the
three examples dealt with in this contribution would look like in the current
formalism, see Examples 2 and 3 of section 6 of Chapter 5 and Example 5 of
section 5 of Chapter 6. At this workshop we promoted for the first time the
idea to assume unique identification of messages in order to achieve a simple
and elegant specification of message passing systems in temporal logic. At
that time we were criticized for introducing such an assumption. Several
years later (in [Koy 87]) we defended ourselves and showed not only that
such a simple temporal logic specification could only be given under this assumption (the alternative is to use much stronger logics), but also that this
assumption was not as restrictive as it may look at first sight (see Example 1
of section 6 of Chapter 5 for the specification of pure message passing systems
in the current formalism). [KKZ 87] again demonstrated the possibility to
apply the special temporal logic to specify message passing and time-critical
systems (see Examples 2 and 4 of section 5 of Chapter 6 for the specification
in the current formalism). In this paper the logic was refined again: now the
state sequence and time were completely decoupled. This is a more faithful
representation of *real* time in real-time systems: the wall clock progresses
independently from the system's execution. The quantitative operators were
defined by the two additional operations of addition and subtraction on time.
In [KKZ 88] a study was made of the fundamentals of real-time by means of
a classification of real-time systems by presence or absence of certain characteristics and several paradigms of real-time systems were given. That paper
also contained an initial and informal overview of requirements for specification languages for real-time. This was subsequently worked out in a formal
framework in [KKZ 89]. In the meantime, the foundations of the special
temporal logic were reexamined which led to two new ideas. The first idea
took the quantitative element (represented by addition and subtraction in
the time domain itself) out of the time domain by adding a distance function that indicates how far two points in time are apart. The range of this
distance function is called the metric domain. The advantage of this representation is its flexibility: instead of measuring time in the time domain
itself different choices for the metric domain provide different possibilities for
measuring time. Furthermore, addition and subtraction do not always provide the means to define the distance function completely. Corresponding to
this idea of posing a metric on time, the special temporal logic was renamed
metric temporal logic. The specification of real-time properties with metric

temporal logic is the subject of [Koy 90]. The second idea emerged from our wish to separate qualitative and quantitative timing aspects already in our new temporal models (including apart from an order on time also a distance function). Since we allowed a pure qualitative view on time (only involving the order), it seemed natural to allow also a purely quantitative view, only involving the distance function. From a semantic point of view, quantifying the metric elements away in operators combining the order and the distance function, gave back the purely qualitative operators of standard temporal logic. Applying the same to pure metric operators lead to the operators **A** (i.e. at every point in time) and **E** (i.e. at some point in time). From there it was only a small step to the irreflexive version of **E**, the **D**-operator which proved to be very versatile.

As stated in Chapter 1, the main objective of this monograph was to develop a specification method for message passing and time-critical systems. As was also mentioned there, the development of such a method should go hand in hand with checking whether the resulting theory really works in practice. In this respect this monograph on one hand incorporates pure fundamental studies (such as Chapter 4, section 4 of Chapter 5 and section 4 of Chapter 6) and on the other hand aims at real applications in practice as is witnessed by the specification examples in section 6 of Chapter 5 and section 5 of Chapter 6. When applying the theory in these specification examples we strived more for clarity than for utmost formality: in most cases we presented informal arguments reasoning on an intuitive semantical level. However, this intuition corresponds exactly to the semantics of the temporal logics involved so that the presented arguments can readily be transformed into rigorous proofs in a straightforward way. In some cases this has been demonstrated.

In this monograph we did not consider several relevant and closely related issues of which we mention a few now. With respect to the application area, as already said, real-time systems exhibit many more features besides that of time-critical aspects, such as reliability, safety and performance. Part of these are covered by the current method, since these topics cannot be treated independently from time-critical aspects, for instance the coupling between response time and performance. As for the specification method, we did not pay much attention to the verification aspect in all its formal detail (as stated above, most of our reasoning was done on a semantical instead of a proof theoretical level). Also hierarchical development was not treated in depth (it featured only in Example 3 of section 6 of Chapter 5). We envisage

that such topics can be treated more extensively on the same footing as
was done for standard temporal logic (see e.g. [MP 82],[MP 83a],[MP 83b]
for verification methods and e.g. [Lam 83b],[BK 85a],[BK 85b],[BKP 84] for
hierarchical development).

As far as we did not do so already in the previous chapters, we now look
at some related work. Formal methods for message passing systems have
been around for some time. For example, [MCS 82] describes safety and liveness properties of message passing networks by a hierarchical method based
upon a compositional specification method for component processes, [SS 82]
uses inference rules for proving partial correctness of concurrent programs
that use message passing for synchronization and communication, [SM 82]
compares specification languages for communication protocols and [HO 83]
treats modular verification of such protocols.

Concerning real-time systems, a review of formal methods for describing
these systems is given in [JG 88]. As far as we know, [BH 81] was the first
paper (using temporal logic) to specify timing characteristics of real-time
systems formally. Their approach differs at several points from ours. Firstly,
they use only real-time operators related to temporal implication instead of
the more powerful operators of metric temporal logic. Secondly, they use the
interleaving model. Consequently their method is restricted to uniprocessor
implementations. Thirdly, their method is limited to specific safety properties. [PH 88] contains a brief account of some attempts to use temporal logic
for the specification of real-time systems. The computational model used is
a timed interleaving model where enabled transitions have associated lower
and upper bounds within which they must be taken. It considers two possible
extensions of temporal logic to deal with real-time. The first adds a global
clock as an explicit variable to which the specification may refer. The second
approach introduces quantitative temporal operators and is very much akin
to metric temporal logic. For specifying synchronous systems it recommends
the use of a discrete time domain (such as the natural numbers) and for asynchronous systems a dense time domain (such as the rationals). One of the
methods using the first approach is that of Ostroff ([Ost 87],[Ost 89]). It introduces a distinguished variable t representing the clock. A typical formula
of his logic RTTL (Real-Time Temporal Logic) is the following:

$$\varphi \wedge t = T \ \rightarrow \ \Diamond (\psi \wedge t \leq T + 5)$$

where T is a global variable.

The semantics of this formula corresponds to the MTL formula

$$\varphi \;\rightarrow\; \Diamond_{\leq 5}\,\psi.$$

As is obvious from this example, metric temporal logic provides a more concise and natural way of specifying real-time properties: the explicit clock variable is against the original philosophy of temporal logic to abstract from time as much as possible (and in the case of real-time it is sufficient to add only terms for expressing time units as in the MTL formula above). On the other hand, RTTL is based on the work of Manna and Pnueli (see the Bibliography) and a sound proof system based on their work is immediately available. An example using the second approach is [GMM 89] incorporating an executable specification language. Another formal approach to the specification of real-time systems, not based on temporal logic, is the Real-Time Logic (RTL) of Jahanian and Mok ([JM 86],[JM 87]). Events are central in RTL and reasoning about real-time systems is based on assertions about the occurrences of events which are mapped by the 'occurrence function' into the time domain of the natural numbers. The use of RTL is restricted to the specification of safety properties.

Complexity and expressiveness results for real-time temporal logics have been investigated in recent work by Alur and Henzinger (see in particular [AH 89],[AH 90],[Hen 91]). In [AH 89] the real-time temporal logic TPTL is introduced. Its distinguished feature is the introduction of a restricted form of quantification, called temporal quantification, which binds a variable to the time(s) it refers to. The temporal operators function then as quantifiers over time variables. For instance, the typical promptness requirement that every p is followed by a q within 5 time units (cf. the similar property specified in RTTL and MTL above) is specified in TPTL by

$$\Box\,x.(p \;\rightarrow\; \Diamond\,y.(q \,\wedge\, y \leq x + 5)).$$

The operators $\Box\,x.\,\varphi$ and $\Diamond\,x.\,\varphi$ can be viewed as restricted quantifiers in a first-order temporal logic like RTTL. Their meaning then translates into $\Box\,\forall\,x\,(t = x \;\rightarrow\; \varphi)$, respectively $\Diamond\,\exists\,x\,(t = x \,\wedge\, \varphi)$.

[AH 90] uses the framework of timed state sequences (a state sequence combined with a unary monotonic function that maps every state to its time) to classify a wide variety of real-time temporal logics according to their complexity and expressiveness. Most formalisms proposed in the literature turn out to be undecidable. [AH 90] identifies two elementary real-time temporal logics as expressively complete fragments of the theory of timed state sequences, and gives tableau-based decision procedures. The logics in question are TPTL and metric temporal logic, both interpreted over timed state sequences (using the natural numbers as the underlying time domain). It is also shown that these two logics are equally expressive. They conclude that these two formalisms are well-suited for the specification and verification of real-time systems.

In [HLP 90] a decidable fragment of RTTL, called XCTL, is proposed where quantification over global time variables is restricted to the outermost level. Timing expressions are rather general though, for example, addition over the time domain is allowed. [HLP 90] shows that the expressive power of XCTL and TPTL is incomparable. Furthermore, XCTL has a singly exponential tableau-based decision procedure and a doubly exponential model checking algorithm.

Also worth mentioning is a recent paper ([AFH 91]) in which a metric interval temporal logic (MITL) is defined. MITL is interpreted over a dense time domain but it is still decidable. Careful analysis shows that decidability is endangered by timing properties of the form

$$\Box \, (p \, \rightarrow \, \mathbf{F}_5 \, q).$$

MITL syntactically prohibits such properties which have infinite timing accuracy. Instead MITL can constrain time differences only with finite (yet arbitrary) precision. For example, the property above is not allowed, but only an approximation such as

$$\Box \, (p \, \rightarrow \, \mathbf{F}_{(4.9, 5.1)} \, q).$$

[AFH 91] shows decidability of MITL in EXPSPACE and uses this result to develop a EXPSPACE-complete model checking algorithm.

The effectiveness of a formal verification technique in practice is greatly influenced by the decidability of the underlying logic. In [Hen 91] several causes for undecidability are investigated. First of all the choice of the time domain is crucial. Whenever logics with a MTL-like syntax (the bounded operators approach) are interpreted over a dense time domain they become undecidable. This also includes branching-time logics using the bounded operators approach such as those considered in [ACD 90] and [Lew 90]. An exception is MITL as has been shown above. As a second source of undecidability, the allowed operations on time are important. [Hen 91] shows that addition of time variables is already sufficient to get undecidability. An exception in this case is XCTL as introduced above. This is accomplished, however, by a severe restriction on the use of quantification over time variables. On the other hand, when timing constraints are restricted to comparison, successor and congruence operations on time, decidability can be obtained. This is illustrated in [Hen 91] in the cases of TPTL and MTL.

Above we mentioned two styles for specifying real-time properties with temporal logic: the explicit clock approach as in RTTL and the bounded operators approach as in MTL. In [HMP 91] two very different proof methodologies for verification expressed in these styles are presented and compared. For the explicit clock approach a relatively (to state reasoning) complete proof system for proving bounded invariance and bounded response properties is given. For the bounded operator approach only a partial result in this direction is obtained. The style of the given proof methodologies is quite different: the bounded operators approach uses complex proof lattices for liveness-like properties and simple local invariants while the explicit clock approach employs a simple safety rule but needs a powerful global invariant.

As to directions for future research, the ideas underlying Chapter 4 have been developed rather recently and many interesting questions remain. To mention a few topics: the exact expressive power of the logics with inequality (e.g. obtained by a precise characterization in correspondence theory), decision procedures, general completeness results for frames and axiomatizations of special structures. Concerning Chapter 5 it would be interesting to find for each class of message passing systems a temporal logic that is sufficient to specify merely this class. In this way one would get a correspondence between certain properties of message passing systems and the essential ingredients needed for (reasoning about) their temporal formalization. Regarding Chapter 6, one of the main remaining questions there is to find a suitable subset of metric temporal logic with a complete axiomatization (and preferably decidable) in order to get an associated verification theory (and possibly even mechanical assistance from a decision procedure). Furthermore, it remains to be seen how we can apply metric temporal logic to medium and large scale examples. Before this can be done it must be sorted out how we can embed such a specification formalism into a method that supports hierarchical development and caters for the description of complex data structures.

A first step in this direction, incorporating metric temporal logic into a hierarchical development method, is provided in [Hoo 91] where MTL is used as the assertion language for correctness formulae of the form P **sat** φ. The intuition behind such a correctness formula P **sat** φ is that it expresses that program P satisfies (MTL) formula φ, for instance the fact that a program P communicates along channel c within 25 time units can be expressed by

$$P \text{ \textbf{sat} } \diamondsuit_{<25} \, comm(c).$$

To obtain a temporal logic axiomatization of the programming constructs of sequential composition and iteration metric temporal logic is extended with operators similar to the chop-operators $\mathcal{C}$ and $\mathcal{C}^*$ from [BKP 84]. [Hoo 91] goes on to formulate a compositional proof system based on correctness formulae of the form P **sat** φ for a simple real-time programming language akin to Occam. The given proof system is sound and relatively complete (relatively with respect to provability of all valid MTL formulae). Later on this proof system is extended for a programming language incorporating program variables and multiprogramming (i.e. sharing of processors).

Bibliography

[ACD 90]   R. Alur, C. Courcoubetis, D. Dill. *Model Checking for Real-Time Systems,* Proceedings of the Fifth IEEE Symposium on Logic in Computer Science, pp. 414–425, 1990.

[Ack 62]   W. Ackermann. *Solvable Cases of the Decision Problem.* Studies in Logic and the Foundations of Mathematics, North-Holland, Amsterdam, 1962.

[Ada 83]   *The Programming Language Ada, Reference Manual.* Lecture Notes in Computer Science Vol. 155, Springer, Berlin, 1983.

[AFH 91]   R. Alur, T. Feder, T.A. Henzinger. *The Benefits of Relaxing Punctuality,* to appear in Proceedings of the Tenth ACM Symposium on Principles of Distributed Computing, 1991.

[AH 89]   R. Alur, T.A. Henzinger. *A Really Temporal Logic,* Proceedings of the Thirtieth Symposium on the Foundations of Computer Science, pp. 164–169, 1989.

[AH 90]   R. Alur, T.A. Henzinger. *Real-time Logics: Complexity and Expressiveness,* Proceedings of the Fifth IEEE Symposium on Logic in Computer Science, pp. 390–401, 1990.

[BC 85]   G. Berry, L. Cosserat. *The ESTEREL Synchronous Programming Language and its Mathematical Semantics,* pp. 389–448 in Proceedings of a *Seminar on Concurrency,* Carnegie Mellon University, Pittsburgh, July 1984, Lecture Notes in Computer Science Vol. 197, Springer, Berlin, 1985.

[Ben 83]   J.F.A.K. van Benthem. *The Logic of Time.* Reidel, Dordrecht, 1983 (second edition 1991).

[Ben 84]   J.F.A.K. van Benthem. *Correspondence Theory*, in [GG 84], pp. 167–247.

[Ben 85]   J.F.A.K. van Benthem. *Modal Logic and Classical Logic*. Bibliopolis, Naples, 1985.

[Ben 89]   J.F.A.K. van Benthem, *private correspondence*, 1989.

[BH 81]   A. Bernstein, P.K. Harter, Jr. *Proving Real-Time Properties of Programs with Temporal Logic,* Proceedings of the Eighth ACM Symposium on Operating System Principles, pp. 1–11, 1981.

[BK 85a]   H. Barringer, R. Kuiper. *Hierarchical Development of Concurrent Systems in a Temporal Logic Framework,* pp. 35–61 in Proceedings of a *Seminar on Concurrency,* Carnegie Mellon University, Pittsburgh, July 1984, Lecture Notes in Computer Science Vol. 197, Springer, Berlin, 1985.

[BK 85b]   H. Barringer, R. Kuiper. *Towards the Hierarchical, Temporal Logic Specification of Concurrent Systems,* in [DHJR 85], pp. 156–183.

[BKP 84]   H. Barringer, R. Kuiper, A. Pnueli. *Now You May Compose Temporal Logic Specifications,* Proceedings of the Sixteenth ACM Symposium on the Theory of Computing, pp. 51–63, 1984.

[BKP 86]   H. Barringer, R. Kuiper, A. Pnueli. *A Really Abstract Concurrent Model and its Temporal Logic,* Proceedings of the Thirteenth ACM Symposium on the Principles of Programming Languages, pp. 173–183, 1986.

[Bla 89]   P. Blackburn. *Nominal Tense Logic*. Centre for Cognitive Science, University of Edinburgh, 1989.

[Bla 90]   P. Blackburn. *Nominal Tense Logic and Other Sorted Intensional Frameworks,* Ph.D. Thesis, University of Edinburgh, 1990.

[Bur 84]   J.P. Burgess. *Basic Tense Logic*, in [GG 84], pp. 89–133.

[CHILL 80]   *CHILL Recommendation Z.200 (CHILL Language Definition)*. C.C.I.T.T. Study Group XI, 1980.

[CK 73]   C.C. Chang, H.J. Keisler. *Model Theory,* Studies in Logic and the Foundation of Mathematics Vol. 73, North-Holland, Amsterdam, 1973.

[Coc 84]   N.B. Cocchiarella. *Philosophical Perspectives on Quantification in Tense and Modal Logic,* in [GG 84], pp. 309–353.

[DHJR 85]   T. Denvir, W. Harwood, M. Jackson, M. Ray. *The Analysis of Concurrent Systems.* Proceedings of a Tutorial and Workshop, Cambridge University, September 1983, Lecture Notes in Computer Science Vol. 207, Springer, Berlin, 1985.

[Gab 81]   D. Gabbay. *Expressive Functional Completeness in Tense Logic,* pp. 91–117 in U. Mönnich (ed.) *Aspects of Philosophical Logic,* Reidel, Dordrecht, 1981.

[Gar 84]   J.W. Garson. *Quantification in Modal Logic,* in [GG 84], pp. 249–307.

[GG 84]   D. Gabbay, F. Guenthner (eds.). *Handbook of Philosophical Logic, Vol. II.* Reidel, Dordrecht, 1984.

[GG 89]   G. Gargov, V. Goranko. *Modal Logic with Names I.* Linguistic Modelling Laboratory, CICT, Bulgarian Academy of Sciences, Sofia, Bulgaria, and Section of Logic, Faculty of Mathematics and Computer Science, Sofia University, Bulgaria, 1989.

[GMM 89]   C. Ghezzi, D. Mandrioli, A. Morzenti. *TRIO: A Logic Language for Executable Specifications of Real-Time Systems.* Report 89-006, Dipartimento di Elettronica, Politecnico di Milano, 1989.

[Gol 82]   R. Goldblatt. *Axiomatising the Logic of Computer Programming.* Lecture Notes in Computer Science Vol. 130, Springer, Berlin, 1982.

[Gor 88]   V. Goranko. *Modal Definability in Enriched Languages.* Section of Logic, Faculty of Mathematics, Sofia University, Bulgaria, 1988, also appeared in Notre Dame Journal of Formal Logic, Volume 31, Number 1, pp. 81–105, 1990.

[GPSS 80]   D. Gabbay, A. Pnueli, S. Shelah, J. Stavi. *On the Temporal Analysis of Fairness,* Proceedings of the Seventh ACM Symposium on the Principles of Programming Languages, pp. 163–173, 1980.

[Hai 80]   B.T. Hailpern. *Verifying Concurrent Processes Using Temporal Logic.* Ph.D. Thesis, Stanford University, 1980.

[Har 84]   D. Harel. *Dynamic Logic,* in [GG 84], pp. 497–604.

[Har 87]   D. Harel. *Statecharts: A Visual Formalism for Complex Systems,* Science of Computer Programming **8**, pp. 231–274, 1987.

[Hen 91]   T.A. Henzinger. *The Temporal Specification and Verification of Real-Time Systems.* Ph.D. Thesis, Department of Computer Science, Stanford University, 1991.

[HLP 90]   E. Harel, O. Lichtenstein, A. Pnueli. *Explicit Clock Temporal Logic,* Proceedings of the Fifth IEEE Symposium on Logic in Computer Science, pp. 402–413, 1990.

[HMP 91]   T.A. Henzinger, Z. Manna, A. Pnueli. *Temporal Proof Methodologies for Real-time Systems,* Proceedings of the Eighteenth ACM Symposium on the Principles of Programming Languages, 1991.

[HO 83]   B.T. Hailpern, S.S. Owicki. *Modular Verification of Computer Communication Protocols,* IEEE Transactions on Communications, Vol. COM-31, No. 1, pp. 56–68, 1983.

[Hoo 91]   J. Hooman. *Specification and Compositional Verification of Real-Time Systems,* Ph.D. Thesis, Eindhoven University of Technology, 1991.

[Hui 91]   C. Huizing. *Semantics of Reactive Systems: Comparison and Full Abstraction,* Ph.D. Thesis, Eindhoven University of Technology, 1991.

[HW 89]   J. Hooman, J. Widom. *A Temporal-Logic Based Compositional Proof System for Real-Time Message Passing,* Proceedings of the Conference on Parallel Architectures and Languages Europe (PARLE) '89, Eindhoven, 12–16 June 1989, Lecture Notes in Computer Science Vol. 366, pp. 424–441, Springer, Berlin, 1989.

[JG 88]   M. Joseph, A. Goswami. *Formal Description of Real-Time Systems: A Review.* Research Report RR129, Department of Computer Science, University of Warwick, 1988.

[JM 86]   F. Jahanian, A.K. Mok. *Safety Analysis of Timing Properties in Real-Time Systems,* IEEE Transactions on Software Engineering 12, pp. 890–904, 1986.

[JM 87]   F. Jahanian, A.K. Mok. *A Graph-Theoretic Approach for Timing Analysis and its Implementation,* IEEE Transactions on Computers, Vol. C-36, No. 8, pp. 961–975, 1987.

[Jon 87]   B. Jonsson. *Compositional Verification of Distributed Systems.* Ph.D. Thesis, Department of Computer Systems, Uppsala University, 1987.

[Kam 68]   J.A.W. Kamp. *Tense Logic and the Theory of Linear Order.* Ph.D. Thesis, University of California, Los Angeles, 1968.

[KKZ 87]   R. Koymans, R. Kuiper, E. Zijlstra. *Specifying Message Passing and Real-Time Systems with Real-Time Temporal Logic,* Proceedings of the Fourth ESPRIT Conference, pp. 311–324, North-Holland, Amsterdam, 1987.

[KKZ 88]   R. Koymans, R. Kuiper, E. Zijlstra. *Paradigms for Real-Time Systems,* Proceedings of a Symposium on Formal Techniques in Real-Time and Fault-Tolerant Systems (ed. M. Joseph), University of Warwick, 22–23 September 1988, Lecture Notes in Computer Science Vol. 331, pp. 159–174, Springer, Berlin, 1988.

[KKZ 89]   R. Koymans, R. Kuiper, E. Zijlstra. *Specification Specified,* chapter 1 of R. Kuiper, *Combining Linear Time Temporal Logic Descriptions of Concurrent Computations,* Ph.D. Thesis, Eindhoven University of Technology, 1989.

[Koy 87]   R. Koymans. *Specifying Message Passing Systems Requires Extending Temporal Logic,* Proceedings of the Sixth ACM Symposium on Principles of Distributed Computing, pp. 191–204,

1987, updated version appeared in Proceedings of a Colloquium on Temporal Logic in Specification, University of Manchester, April 1987, Lecture Notes in Computer Science Vol. 398, pp. 213–223, Springer, Berlin, 1989.

[Koy 89]   R. Koymans. *Specifying Message Passing and Time-Critical Systems with Temporal Logic,* Ph.D. Thesis, Eindhoven University of Technology, 1989.

[Koy 90]   R. Koymans. *Specifying Real-Time Properties with Metric Temporal Logic,* Journal of Real-Time Systems, Volume 2, Number 4, pp. 255–299, Kluwer Academic Publishers, 1990.

[KP 87]   S. Katz, D. Peled. *Interleaving Set Temporal Logic,* Proceedings of the Sixth ACM Symposium on Principles of Distributed Computing, pp. 178–190, 1987.

[KR 85]   R. Koymans, W.-P. de Roever. *Examples of a Real-Time Temporal Logic Specification,* in [DHJR 85], pp. 231–251.

[KSRGA 85] R. Koymans, R.K. Shyamasundar, W.-P. de Roever, R. Gerth, S. Arun-Kumar. *Compositional Semantics for Real-Time Distributed Computing,* Proceedings of the Workshop on Logics of Programs '85, Lecture Notes in Computer Science Vol. 193, pp. 167–189, Springer, Berlin, 1985, extended version appeared in Information and Computation, Volume 79, Number 3, pp. 210–256, Academic Press, December 1988.

[KVR 83]   R. Koymans, J. Vytopil, W.-P. de Roever. *Real-Time Programming and Asynchronous Message Passing,* Proceedings of the Second ACM Symposium on Principles of Distributed Computing, pp. 187–197, 1983.

[Lam 83a]   L. Lamport. *What Good is Temporal Logic?,* Proceedings of Information Processing (IFIP) '83 (editor R. Mason), pp. 657–668, North-Holland, Amsterdam, 1983.

[Lam 83b]   L. Lamport. *Specifying Concurrent Program Modules,* ACM Transactions on Programming Languages and Systems (TOPLAS), Volume 5, Number 2, pp. 190–223, 1983.

[Lam 85]   L. Lamport. *STL/SERC Problems,* in [DHJR 85], pp. 252–270.

[Lew 90]   H. Lewis. *A Logic of Concrete Time Intervals*, Proceedings of the Fifth IEEE Symposium on Logic in Computer Science, pp. 380–389, 1990.

[LPZ 85]   O. Lichtenstein, A. Pnueli, L. Zuck. *The Glory of The Past*, Proceedings of the Workshop on Logics of Programs '85, Lecture Notes in Computer Science Vol. 193, pp. 196–218, Springer, Berlin, 1985.

[MCS 82]   J. Misra, K.M. Chandy, T. Smith. *Proving Safety and Liveness of Communicating Processes, with Examples,* Proceedings of the First ACM Symposium on Principles of Distributed Computing, 1982.

[MM 84]   B. Moszkowski, Z. Manna. *Reasoning in Interval Temporal Logic,* Proceedings of AMC/NSF/ONR Workshop on Logics of Programs, Lecture Notes in Computer Science Vol. 164, pp. 371–383, Springer, Berlin, 1984.

[Mos 83]   B. Moszkowski. *Reasoning about Digital Circuits.* Ph.D. Thesis, Department of Computer Science, Stanford University, 1983.

[Mos 86]   B. Moszkowski. *Executing Temporal Logic Programs.* Cambridge University Press, 1986.

[MP 82]   Z. Manna, A. Pnueli. *Verification of Concurrent Programs: The Temporal Framework,* pp. 215–273 in R. Boyer, J. Moore (eds.) *The Correctness Problem in Computer Science,* International Lecture Series in Computer Science, Academic Press, London, 1982.

[MP 83a]   Z. Manna, A. Pnueli. *How to Cook a Temporal Proof System for your Pet Language,* Proceedings of the Tenth ACM Symposium on the Principles of Programming Languages, pp. 141–154, 1983.

[MP 83b]   Z. Manna, A. Pnueli. *Verification of Concurrent Programs: A Temporal Proof System,* pp. 163–255 in J. de Bakker, J. van Leeuwen (eds.) *Foundations of Computer Science IV,* Mathematical Center Tracts Vol. 159, CWI, Amsterdam, 1983.

[MP 87]   Z. Manna, A. Pnueli. *A Hierarchy of Temporal Properties.* Department of Computer Science, Stanford University, Report No. STAN-CS-87-1186, October 1987.

[MP 89]   Z. Manna, A. Pnueli. *The Anchored Version of the Temporal Framework,* pp. 201–284 in J.W. de Bakker, W.-P. de Roever, G. Rozenberg (eds.) *Linear Time, Branching Time and Partial Order in Logics and Models for Concurrency,* Lecture Notes in Computer Science Vol. 354, Springer, Berlin, 1989.

[Ost 87]   J.S. Ostroff. *Real-Time Computer Control of Discrete Event Systems Modelled by Extended State Machines: A Temporal Logic Approach.* Ph.D. Thesis, Department of Electrical Engineering, University of Toronto, 1987.

[Ost 89]   J.S. Ostroff. *Temporal Logic for Real-Time Systems.* Advanced Software Development Series, Research Studies Press Limited (marketed by John Wiley & Sons), England, 1989.

[Pan 88]   P.K. Pandya. *Compositional Verification of Distributed Programs.* Ph.D. Thesis, Tata Institute of Fundamental Research, Bombay, 1988.

[Par 81]   D. Park. *Concurrency and Automata on Infinite Sequences,* Proceedings of Fifth GI (Gesellschaft für Informatik) Conference, Lecture Notes in Computer Science Vol. 104, Springer, Berlin, 1981.

[Pen 88]   W. Penczek. *A Temporal Logic for Event Structures,* Fundamenta Informaticae XI, pp. 297–326, 1988.

[PH 88]   A. Pnueli, E. Harel. *Applications of Temporal Logic to the Specification of Real Time Systems,* Proceedings of a Symposium on Formal Techniques in Real-Time and Fault-Tolerant Systems (ed. M. Joseph), University of Warwick, 22–23 September 1988, Lecture Notes in Computer Science Vol. 331, pp. 84–98, Springer, Berlin, 1988.

[Pnu 77]   A. Pnueli. *The Temporal Logic of Programs,* Proceedings of the Eighteenth Symposium on the Foundations of Computer Science, pp. 46–57, 1977.

[Pnu 86]   A. Pnueli. *Applications of Temporal Logic to the Specification and Verification of Reactive Systems: A Survey of Current Trends*, pp. 510–584 in J.W. de Bakker, W.-P. de Roever, G. Rozenberg (eds.) *Current Trends in Concurrency*, Lecture Notes in Computer Science Vol. 224, Springer, Berlin, 1986.

[Pri 67]   A. Prior. *Past, Present and Future*. Oxford University Press, London, 1967.

[PW 84]   S. Pinter, P. Wolper. *A Temporal Logic for Reasoning about Partially Ordered Computations*, Proceedings of the Third ACM Symposium on Principles of Distributed Computing, pp. 28–37, 1984.

[Rijk 89]   M. de Rijke. *The Modal Theory of Inequality*. Master Thesis, Faculty of Mathematics and Computer Science, University of Amsterdam, 1989.

[Rijk 90]   M. de Rijke. *The Modal Theory of Inequality*, updated version of [Rijk 89], ITLI Prepublication Series LP-90-15, Institute for Language, Logic and Information, University of Amsterdam, 1990.

[Sah 75]   H. Sahlqvist. *Completeness and Correspondence in the First and Second Order Semantics for Modal Logic*, pp. 110–143 in S. Kanger (ed.) *Proceedings of the Third Scandinavian Logic Symposium*, North-Holland, Amsterdam, 1975.

[SCFG 82]   A.P. Sistla, E.M. Clarke, N. Francez, Y. Gurevich. *Can Message Buffers Be Characterized in Linear Temporal Logic?*, Proceedings of the First ACM Symposium on Principles of Distributed Computing, pp. 148–156, 1982.

[SCFM 84]   A.P. Sistla, E.M. Clarke, N. Francez, A.R. Meyer. *Can Message Buffers Be Axiomatized in Linear Temporal Logic?*, Information and Control, Volume 63, pp. 88–112, 1984.

[Seg 70]   K. Segerberg. *Modal Logics with Linear Alternative Relations*, Theoria, Volume 36, pp. 301–322, 1970.

[Seg 71]   K. Segerberg. *An Essay in Classical Modal Logic.* Filosofika Studier **13**, Department of Philosophy, University of Uppsala, 1971.

[SM 82]   R.L. Schwartz, P.M. Melliar-Smith. *From State Machines to Temporal Logic: Specification Methods for Protocol Standards,* IEEE Transactions on Communications Vol. COM-30, No. 12, pp. 2486–2496, 1982.

[SMV 83]   R.L. Schwartz, P.M. Melliar-Smith, F.H. Vogt. *An Interval Logic for Higher-Level Temporal Reasoning,* Proceedings of the Second ACM Symposium on Principles of Distributed Computing, pp. 173–186, 1983.

[SS 82]   R.D. Schlichting, F.B. Schneider. *Using Message Passing for Distributed Programming: Proof Rules and Disciplines,* Proceedings of the First ACM Symposium on Principles of Distributed Computing, 1982, also appeared in ACM Transactions on Programming Languages and Systems (TOPLAS), Volume 6, Number 3, pp. 402–431, July 1984.

[Sta 79]   J. Stavi. *Functional Completeness Over the Rationals.* Unpublished, Bar-Ilan University, Ramat-Gan, Israel, 1979.

[Sti 87]   C. Stirling. *Comparing Linear and Branching Time Temporal Logics.* ECS-LFCS-87-24, Laboratory for Foundations of Computer Science, Department of Computer Science, University of Edinburgh, April 1987.

[Tar 69]   A. Tarski. *What is Elementary Geometry?,* pp. 164–175 in J. Hintikka (ed.) *The Philosophy of Mathematics,* Oxford University Press, London, 1969.

[Tho 86]   W. Thomas. *Safety- and Liveness-Properties in Propositional Temporal Logic: Characterizations and Decidability.* Schriften zur Informatik und Angewandten Mathematik, Bericht Nr. 116, Rheinisch-Westfälische Technische Hochschule Aachen, April 1986.

[Wol 86]   P. Wolper. *Expressing Interesting Properties of Programs in Propositional Temporal Logic,* Proceedings of the Thirteenth

ACM Symposium on the Principles of Programming Languages, pp. 184–193, 1986.

[ZRE 85]   J. Zwiers, W.-P. de Roever, P. van Emde Boas. *Compositionality and Concurrent Networks: Soundness and Completeness of a Proofsystem*, Proceedings of the Twelfth International Colloquium on Automata, Languages and Programming, Lecture Notes in Computer Science Vol. 194, pp. 509–519, Springer, Berlin, 1985.

[Zwi 88]   J. Zwiers. *Compositionality, Concurrency and Partial Correctness: Proof Theories for Networks of Processes, and their Connection.* Ph.D. Thesis, Eindhoven University of Technology, 1988, also appeared as Lecture Notes in Computer Science Vol. 321, Springer, Berlin, 1989.

Index

Lecture Notes in Computer Science

For information about Vols. 1–579
please contact your bookseller or Springer-Verlag